The Representation of Monkeys in the Art and Thought of Mediterranean Cultures

A new perspective on ancient primates

Cybelle Greenlaw

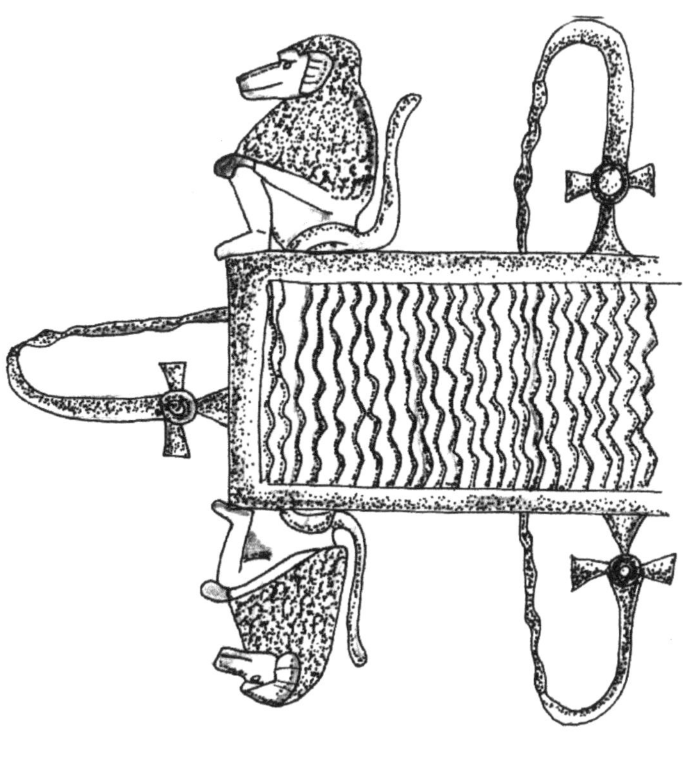

BAR International Series 2192
2011

Published in 2016 by
BAR Publishing, Oxford

BAR International Series 2192

The Representation of Monkeys in the Art and Thought of Mediterranean Cultures

ISBN 978 1 4073 0747 3

BAR Publishing is the trading name of British Archaeological Reports (Oxford) Ltd. British Archaeological Reports was first incorporated in 1974 to publish the BAR Series, International and British. In 1992 Hadrian Books Ltd became part of the BAR group. This volume was originally published by Archaeopress in conjunction with British Archaeological Reports (Oxford) Ltd / Hadrian Books Ltd, the Series principal publisher, in 2011. This present volume is published by BAR Publishing, 2016.

Printed in England

BAR
PUBLISHING

BAR titles are available from:

BAR Publishing
122 Banbury Rd, Oxford, OX2 7BP, UK
EMAIL info@barpublishing.com
PHONE +44 (0)1865 310431
FAX +44 (0)1865 316916
www.barpublishing.com

Table of Contents

List of Illustrations

All drawn illustrations by Stacey Ramey

Acknowledgements

This book, adapted from my PhD thesis, could not have been completed without the support and encouragement of my supervisor, Dr. Christine Morris, my mother, Rosemary Greenlaw, who inspired my study and love of monkeys, and my dear friends in Ireland and the U.S., who sent me every reference to monkeys they encountered, listened to my dilemmas, and helped with proofreading. I also owe a debt of gratitude to all the baboons and macaques I have known and loved over the years. Without them, my life would be much poorer. Special thanks are also owed to my examiners, Dr. Susan Sherratt and Dr. Hazel Dodge, whose comments and criticisms were extremely helpful. Thanks also to Stacey Ramey for her illustrations.

I would also like to acknowledge the support I received from all members of the School of Classics at Trinity College Dublin during my studies there from 2001-2004. Additionally, without the generous Trinity Scholarship, I could not have completed my degree.

My thanks also go to BAR for their patience and support on the road to publication.

INTRODUCTION

People often ask me, "Why monkeys?" Then they ask, "Were there monkeys in the Mediterranean?" In answer to the first question, my interest in monkeys began in 1984, when my mother and I decided to convert our guesthouse into a primate sanctuary for baboons and macaques. After more than a decade of building cages, cleaning cages and digging drainage ditches, I decided to leave the monkeying around to my mother and do a Bachelor's degree in French literature with a minor in Classics. All was going well until my third year as an undergraduate when I saw the blue monkeys of Thera—I knew there would never be any way out of the monkey business.

Over the years, I have come to realize that people almost always have a strong reaction to non-human primates. Some people praise their beauty, sensitivity and intelligence, while others mock their ugliness, promiscuity and stupidity. The same reactions are apparent in ancient Mediterranean cultures. The Egyptians and Minoans valued them highly. The Greeks and Romans were far less enthusiastic. In her essay on feminism and functionalism in primate studies entitled "Baboons with Briefcases vs. Langurs in Lipstick," Sperling comments, "Primates are icons for us. They seem to live at the boundary of nature and culture, and the ways they appear in current Western symbolism reflect the political and socioeconomic discourses of the historical periods during which primate studies has developed as a discipline" (1991: 222). I would take the statement much further—primates have always been icons for humans. They are the animals that most closely resemble us, and how any given society interprets their behaviour and appearance reveals much more about its own cultural norms than about primates.

In answer to the second question, it is hardly surprising that so many people are unaware of all the monkeys in the ancient Mediterranean. No major work has been written on the subject since William Coffman McDermott's *The Ape in Antiquity*, published in 1938. McDermott is still the most commonly cited authority on ancient monkeys, and his book remains an excellent source for references to non-human primates in Greek and Latin literature. It even provides a useful catalogue of simian iconography in a variety of media.

However, many of McDermott's comments on primate behaviour and attempts to identify the species are outdated and highly inaccurate. Unfortunately, the majority of more recent books and articles have failed to question his knowledge of these areas, and a few scholars have even created new and more outlandish inaccuracies. Toynbee, for example, seems to have been confused by ancient references to monkeys, stating:

> The breed of monkeys best known in Italy was the tailed Barbary ape (*cercopithecus*), of which there is also a tailless variety. Pliny states that it came from Ethiopia and had a black head, hair like that of a onkey, and a voice quite different from that of other monkeys (1973: 56).

Barbaries are, in fact, macaques, belonging to the genus *Macaca*. They are not apes or related to the monkeys currently classified as *Cercopitheci*. Only a tailless, North African variety exists. The present study aims to explain and amend some of these problems and provide a survey of the diverse cultural attitudes toward monkeys through an examination of the iconographical, physical and textual evidence from several Mediterranean cultures.

Part of McDermott's difficulty, and that of subsequent scholars, in correctly identifying the various species depicted in Mediterranean art may, in fact, be linked to the title of his book: *The* Ape *in Antiquity*. He explains his use of the term in this way:

> ...the word "ape" is used as a general term and as a term where the reference is to the tail-less animal; the word "monkey" has been used in general for apes with tails; more restricted words, such as "langur," "gibbon," "guenon" and so forth, are used only when speaking in particular places where references to apes in ancient times seems to indicate these generic names (1938: 102).

Now, of course, the word 'ape' is used specifically to describe a small number of species from the order of primates, namely the Hylobates or lesser apes, the gibbons and siamangs, and the Pongidae or great apes, which include the gorilla (*Gorilla gorilla*), the chimpanzee and bonobo (*Pan troglodytes* and *Pan paniscus*) and the orang-utan (*Pongo pygmaeus*). It does not include any other primate, except perhaps human beings. Some primatologists would prefer to place the genera *Gorilla* and *Pan* under the family Hominidae, which currently comprises a single species, *Homo sapiens* (Nowak 1999: 186-7).

Clearly, far more research has been conducted on primate behaviour, intellectual capabilities and the range of their natural habitats since 1938, and many of the Old World primates discussed in McDermott's book have been reclassified and assigned new binomials. This introduction, then, provides a brief survey of the primate world as we know it at present. Special emphasis is placed on the sub-family, Cercopithecinae, since these are the non-human primates that would have been best known to the peoples inhabiting the regions examined in this book. Additionally, emphasis is placed on the males because they feature more prominently in ancient art. The reasoning behind the need for this introduction is that no monkeys were native to Europe, Crete, the Cycladic Islands, Anatolia or Mesopotamia in historic times. It is unlikely that more than one species was ever native to Egypt. Therefore, with the one possible exception, all depictions of monkeys in the art of these ancient cultures represent imported, exotic animals. Having some background information about the types of monkeys that could easily have been imported to those areas aids in the process of identifying the individual species depicted in iconography from as early as the Old Kingdom and described in much later Greco-Roman medical texts. Some scholars have mistakenly identified representations of Old World monkeys as apes or described them as having certain attributes that could only be present in New World monkeys. This invariably leads to confusion about the types of animals with which the people of these regions were familiar and creates a distorted view of ancient trade routes. Hopefully, this brief survey will lead to a clarification both of the language needed to describe primates and of the species that could have been imported.

Primates can be divided into six main groups—the lemurs; the lorises and galagos; the tarsiers; the New World monkeys; the Old World monkeys; and apes and humans. The first three groups are classified as prosimians. These are simple species of small primates that may be thought of as 'almost monkeys.' Lemurs inhabit Madagascar, while the lorises and galagos are found in Asia and Africa. Many prosimians are nocturnal and all but the tarsiers have a *tapetum lucidum*, a crystalline shield that lies behind the retina, for better night vision. Evidence that the prosimians share a common nocturnal ancestry is supported by their wet noses, a feature characteristic of nocturnal animals requiring a keen sense of smell as compensation for the limitations of night vision (Kavanagh 1983: 26). Some primatologists consider the tarsier's dry nose to be an evolutionary step between prosimians and true simians or monkeys (Preston-Mafham 1999: 9). No prosimians were depicted by any of the cultures considered in this study.

New World monkeys are those simians inhabiting the South American continent, Central America and Mexico. These monkeys are mainly adapted to forest life—no New World counterpart to the African semi-terrestrial baboons and apes exists. Some, such as marmosets and tamarins, are squirrel-like and have claws rather than fingernails. Many of the larger monkeys have prehensile tails that act as fifth limbs. These characteristics are absent from Old World monkeys, which are, of course, the subject of concern here.

Old World monkeys belong to a single family, the Cercopithecidae, and although some controversy exists, it can be divided into at least two sub-families, the Colobinae and the Cercopithecinae, both of which have members in Africa and Asia (Nowak 1999: 129). The first group is composed of leaf-eaters, such as the colobus, which inhabits central and western Africa and the Asian langur. Most of the species are arboreal and have pot-bellied stomachs full of bacteria for breaking down leaves. Although some colobus monkeys can be found as close to Egypt as Ethiopia, their specialized diet would make their survival in a domestic or urban environment difficult.

The Cercopithecinae, on the other hand, are generalists. Kavanagh describes them as "the hustlers of the primate order" (1983: 141). The members of this sub-family have found ways of surviving in a variety of fairly extreme environments and are capable of eating any foods humans can eat. The simians living closest to the aforementioned regions are the hamadryas and anubis baboons (*Papio hamadryas* and *Papio anubis*), patas monkeys and vervets (*Erythrocebus patas* and *Cercopithecus aethiops*), and two species of macaque, the Barbary and the rhesus (*Macaca sylvanus* and *Macaca mulatta*). The baboons and vervets were imported to Egypt even from Predynastic times from the southern lands of Nubia, modern day Ethiopa, and Punt, the exact location of which has yet to be discovered. The patas still has a range extending to the Upper Nile Valley and may have been native to Egypt. The Barbary macaque, the only African macaque, which is most commonly but inaccurately recognized as the ape of Gibraltar, was imported to many sites around the Mediterranean in the 5th century BCE. The species is now only native to Morocco and Algeria, but a colony once existed near Carthage, in modern-day Tunisia. Although macaques could have been introduced to the Rock in ancient times, the earliest record of their habitation there dates to 711 CE. The British Garrison probably established the current population in the 1740's. The popular belief that they would lose the island if the macaques ever vanished from it led them to replenish the population on three separate occasions with monkeys from Morocco when major die offs were caused by parasites, improper feeding practices, and illnesses carried by tourists (Fa and Lind 1996: 235-262).

Mummified specimens of all five species discussed so far have been discovered in sites around Egypt. For the most part, these date from the Late to Ptolemaic Periods, but a few date from the New Kingdom. The third volume of Lortet and Gaillard's study, *La Faune Momifiée de l'Ancienne Égypte*, published in 1907, contains a study of the monkey necropolis at Thebes and is still one of the most commonly cited sources for information on the identification of mummified monkeys. Anderson's 1902 work, *The Zoology of Egypt*, also presents good illustrations and early radiographs of mummified monkeys. More recent studies of primate mummies are discussed in my first chapter. The sixth species, the rhesus macaque, whose westernmost habitat is the area corresponding to modern-day Afghanistan, is the Asian species most likely to have been imported to Mesopotamia. There is no evidence, however, that it was known in Egypt or the western Mediterranean.

To better understand the depictions and references to primates discussed in this study, the six species must be considered in more detail. The hamadryas baboon is one of the smaller, ground-dwelling species of baboon, which inhabits both coasts of the Red Sea. In modern times, its African habitat extends north to south roughly from Suakin, near Port Sudan, to Webi Shebelle River in Somalia (Kummer 1995: 8). Although there has been some controversy, it is unlikely that the hamadryas ever inhabited the lower Nile Valley (Kummer 1995: 3). The species is striking for its pink face and bright red hindquarters. Adult males are silver with a thick cape; females are considerably smaller, brown, and lack the cape (**Fig. 1**). The anubis, on the other hand, is a large

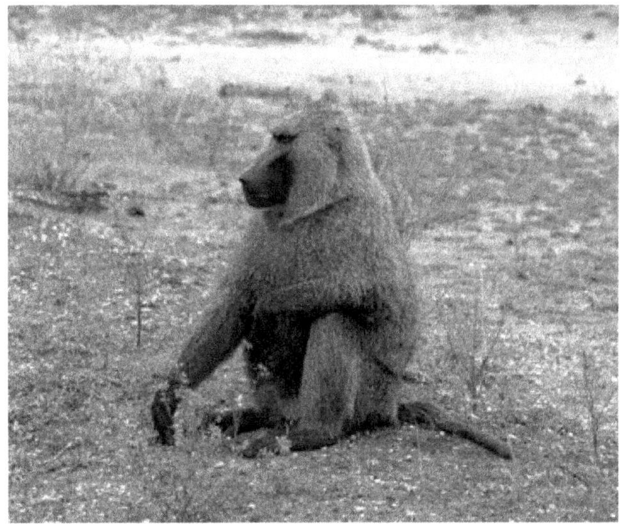

Fig. 2: An Anubis baboon.
Photographer: Benjamin Schalwijk,
courtesy of Dreamstime.com

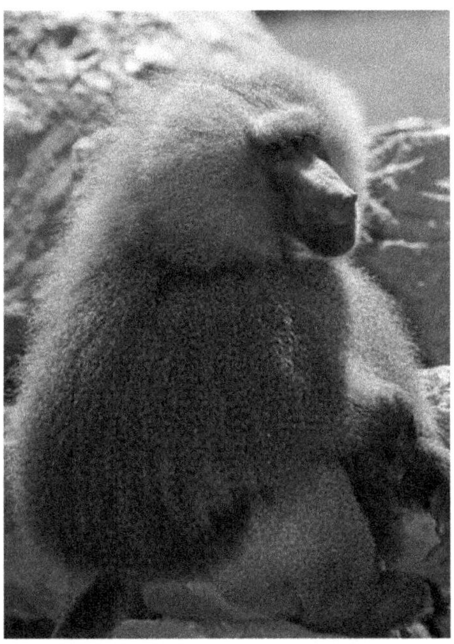

Fig. 1: A male hamadryas baboon.
Photographer: Stephen Meese,
courtesy of Dreamstime.com

Savannah baboon, whose natural range extends inland to south and west of the hamadryas' African habitat. These animals have black faces, a thick olive coat and dark hindquarters (**Fig. 2**). Like the Hamadryas, the male is much larger than the female, but no difference exists in its coloration. Males have a slight cape, but it is in no way comparable to that of the hamadryas.

Although hamadryas and anubis baboons are similar enough to have formed a hybrid zone in Ethiopa, hamadryas society is notably different from that of the Savannah baboons. Male hamadryas baboons have an ability to herd females and have successfully dominated anubis females and brought them into their harems. Anubis males do not establish harems, and the mixed offspring of anubis females also lack this trait (Nowak 1999: 148-50). It seems that this herding of anubis females is a strategy taken by hamadryas males who are not able to attract females from their own troops. This is accomplished by keeping the larger, competing male anubis baboons away from the selected females by means of threats. The female is dominated by warning bites to the neck. When her troop leaves, she is abandoned and compelled to follow her hamadryas kidnapper rather than remain alone to face predators (Kummer 1995: 107-8). Despite this aggressive, violent behaviour, Kummer observed a greater loyalty toward females in hamadryas society. Males will stay with their sick or injured females, even at the risk of separation from their troop, while the anubis baboons tend to abandon females who are not able to follow. In one case, he noticed that a male hamadryas was particularly late returning to the troop's sleeping cliff because he was waiting for his female, who was following slowly on three limbs as she clutched her dead infant in one arm (Kummer 1995: 101). Subadult male hamadryas baboons also try to cultivate families by kidnapping infants. In the wild, the infants are not mistreated and are given back when the father threatens them or the mother 'kecks' at them until her child is returned (Kummer 1995: 213-14). The males are attentive to youngsters in other ways as well. Kummer noticed that a male, who had been watching some youngsters playing on the ground below the cliff where he was sitting, prevented their injury by catching and holding a falling stone that might have hit them (Kummer 1995: 205).

The patas (*Erythrocebus patas*) is a hardy, long-tailed guenon, slightly larger than the vervet, with brick-red fur, a white front and light-coloured face with a distinctive dark marking on its nose and moustache-like fur on its upper lip (**Fig. 3**). Its range covers the Upper Nile Valley to the southern Sahara, west to Senegal and Mauritania (Osborn and Osbornova 1998: 41). For the most part, it occupies the wooded savannah and open grassland but tends to avoid thick forests. It is the fastest primate and has been recorded running 55 km/hr (Nowak 1999: 130).

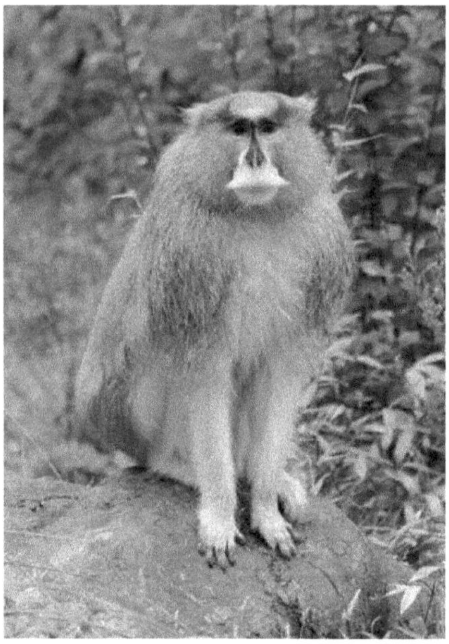

Fig. 3: A patas monkey.
Photographer: Susan Pettitt,
courtesy of Dreamstime.com

The smaller, widely distributed savannah guenons have been the subject of debate in recent years. Some primatologists would differentiate four allopatric species: *Cercopithecus sebaeus*, the green monkey, inhabiting Senegal to the Volta River in Ghana, *C. tantalus*, occupying the zone from the Volta River to Uganda, *C. aethiops*, the grivet, ranging from Sudan to Ethiopia, and *C. pygerythrus*, the vervet, with an enormous habitat ranging from southern Ethiopia to Angola and South Africa. Others consider all of these monkeys to be subspecies of *Cercopithecus aethiops*, and at least one recent authority would reclassify these animals as *Chlorocebus* rather than *Cercopithecus* (Nowak 1999: 132). All four are extremely similar in appearance. Males are a light grey, with brightly coloured blue and red genitalia. Females are greenish-brown, and both sexes have dark faces and hands. All four species have white, fluffy cheek fur, which seems to be most pronounced on the grivet (**Fig. 4**). Because it is practically impossible to distinguish the minor differences even in the most realistic Egyptian depictions and because the terms vervet, grivet and green monkey are often used interchangeably, I have chosen to refer to all such monkeys as vervets.

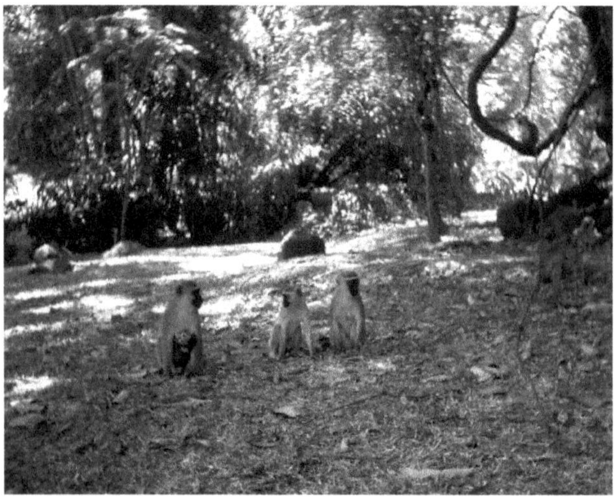

Fig. 4: Vervet monkeys in Zomba, Malawi.
Photo by author

The macaques share many similarities with baboons but are smaller and have much more rounded faces. During their adolescence, male baboons and macaques both begin to grow enormous canine teeth, which continue to grow throughout their lives. The teeth are honed on the lower canines and displayed in open-mouthed threats. Both species yawn when nervous and lip-smack while grooming to demonstrate friendliness or submission. Like baboons, males are considerably larger than females. The macaques, however, are an extremely diverse group of monkeys with much greater variation in their physical features and size than the baboons. Currently, according to Nowak, twenty species of macaques exist in the world, and all but the Barbary have habitats in Asia (1999: 140). This North African macaque is a large monkey without a tail, which is the main reason it is often called an ape. It has light brown fur and light facial skin that becomes tan in the sun (**Fig. 5**). The rhesus macaque has a medium length tail, brown fur and a ruddy face. Its enormous habitat ranges from eastern Afghanistan to India, Nepal, northeastern China and Vietnam, and it has been able to adapt to a variety of extreme climates and elevations (**Fig. 6**; Nowak 1999: 140, 142).

Like the Hanuman langur, it is considered sacred in Hindu religion, and this status provided the animals with a certain amount of protection in areas where the religion was practiced. Now, however, their numbers are dwindling. Urbanization has encroached upon the monkeys' territory, leading troops to raid crops, and the

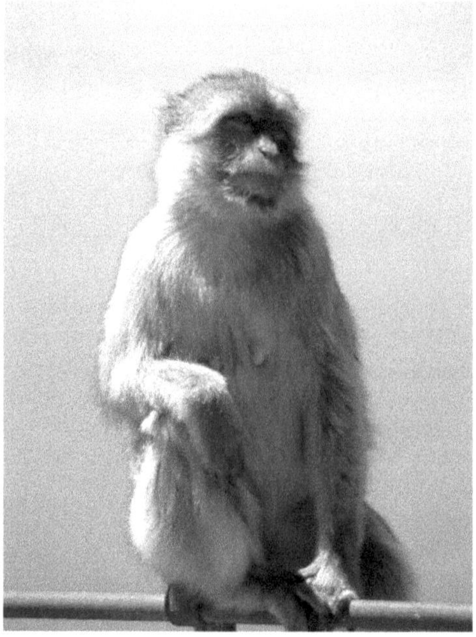

Fig. 5: A Barbary macaque.
Photographer: Tim Seed,
courtesy of freedigitalphotos.net

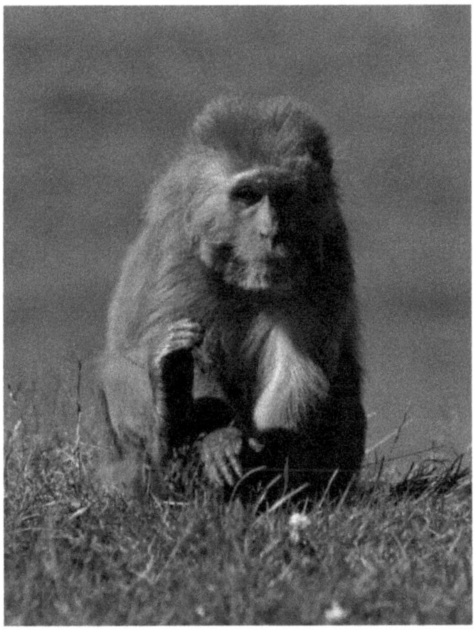

Fig. 6: A rhesus macaque.
Courtesy of freedigitalphotos.net

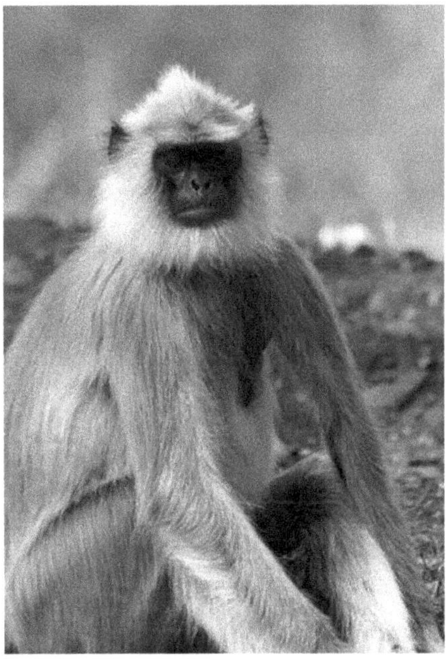

Fig. 7: A Hanuman langur.
Photographer: Nickolay Stanev,
courtesy of Dreamstime.com

animals are increasingly treated as pests. Additionally, the demand for primate research in the United States and other countries had a serious impact on their numbers—as many as 200,000 per year were exported to the US in the 1950's (Nowak 1999: 146). According to Southwick and Siddiqi, from the 1950's to the 1970's, over 1,000,000 rhesus macaques were removed from India for this purpose (1977: 340). The statistics are shocking when one considers that macaques only give birth to one infant at a time, and the gestation period is about six months. This sad fate is, of course, shared by many other non-human primates and, as will be demonstrated in the chapter on Greco-Roman monkeys, is a practice that has its roots in antiquity.

Before concluding this introduction, two other species are worthy of mention. The first is the Hanuman langur, which could also have been introduced to Mesopotamia. Its habitat covers most of India, the foothills of Nepal and the island of Sri Lanka. It is a long-tailed monkey with fur; its bare face, hands and feet are black. The males are 60 percent larger than the females (**Fig. 7**; Kavanagh 1983: 126). Although langurs are Colobinae and have a more specialized diet than the rhesus, they are highly adaptable to a range of climates. The males, like the hamadryas, try to form harems of females, but these monkeys are far more violent. Takeovers are a common occurrence, and infants fathered by the ousted males are routinely killed by the new harem leaders.

The second species that deserves attention is an unusual creature called the gelada, which inhabits Amhara plateau of northern Ethiopa, occasionally crossing paths with hamadryas baboons (**Fig. 8**; Kavanagh 1983: 164). Geladas are similar to baboons, and some authorities have classified them as an aberrant form of the genus Papio. Now, however, most primatologists classify them as the unique living species of the genus *Theropithecus* (Dunbar 1977: 363). The animals have long, shaggy brown fur, and black faces, hands and feet. The males also sport an impressive cape, and the long tufts at the ends of their tails give them a leonine appearance. Both sexes have lumps of red flesh on their chests; the female's caruncles swell to demonstrate her sexuality receptivity. These animals do not seem to have been represented at all in Egyptian art, but as is discussed in Chapter 4, there is some indication that the species was known in Roman times.

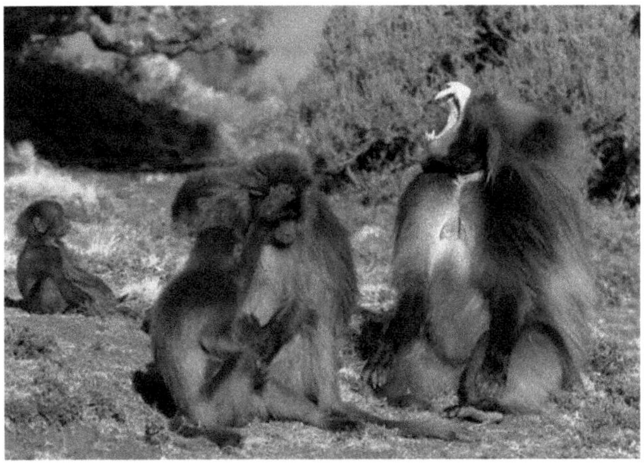

Fig. 8: Geladas.
Photographer: William Davies,
courtesy of Dreamstime.com

Of course, this brief survey cannot cover every physical feature and behavioural trait observed and represented by ancient cultures, but more aspects are considered throughout the study as the need arises. I have concentrated more on the behaviour and characteristics of the males partly because this sexual dichotomy was observed and represented by the Egyptians, who were largely responsible for the spread of simian iconography to other areas of the Mediterranean.

Female monkeys did not represent gods, but they did have a place in society, as is discussed in Chapter 1. The female monkey's role as a caring mother is a common theme in Egyptian iconography and is subsequently repeated in many other cultures. Additionally, the field of primatology had long been dominated by men, and views on primate behaviour were viewed as primitive forms of human male-dominated society. The monkeys were, as Sperling commented, "icons for us." Clearly, this view of primates as dominant males is changing, but to concentrate on the powerful roles of the females would lead to a distorted view of the way in which the animals were observed by ancient peoples. This, in turn, could lead to a distorted view of the attitudes held by the various Mediterranean cultures toward these amazing animals.

Chapter 1
MONKEYS IN EGYPT: FROM THE OLD KINGDOM TO THE PTOLEMAIC PERIOD

From the Old Kingdom to the Ptolemaic Period, non-human primates were assigned numerous functions in Egyptian religion and culture.[1] This chapter will explore their development through an examination of the portrayal of the animals in art and literature. Because monkeys were so much a part of Egyptian culture even from predynastic times, it is not possible to cover all their functions in one chapter. I hope, however, to give a useful and accurate survey of the roles filled by various species and demonstrate the ways in which the Egyptians' perception of the creatures changed over time.

Baboons, in particular, had great religious significance for the Egyptians, while smaller monkeys, such as vervets and patas monkeys, were kept as pets and appeared in a variety of media. During this span of three millennia, representations of these animals and their roles in daily life changed and evolved. Their portrayal as symbols of wealth in tomb chapel scenes from the Old to New Kingdoms grew or diminished according to the changing economic and political environment. New developments in the production of faience led to innovations in the portrayal of the animals, and the growing importance of the Osiris myth may have helped promote the frequent depiction of the baboon as a manifestation of Thoth. However, the use of monkeys in the Old Kingdom must really be seen as a continuation from the late predynastic and early dynastic periods.

Excavations conducted in the early 1900s at Hierakonpolis, Abydos and Elephantine have uncovered numerous glazed composition and limestone figurines, which attest to the existence of early baboon cults (**Figs. 9 and 10**). These were most likely dedicated to Hedjwer, the Great White. This is the earliest identifiable baboon deity, and the name first appears on an ivory label of Semerkhet, dating from c.2900 BCE, found at Abydos (Spencer 1993: 67, fig. 46; T. Wilkinson 1999: 285). A large, alabaster baboon in Berlin (22607), dating from the First Dynasty and bearing the name of King Narmer, may also represent this deity (**Fig. 11**). Hedjwer's significance is not entirely understood, but some scholars have suggested that the early kings may have had themselves portrayed as baboons. Friedman employs an interpretation of the Berlin baboon as a representation of Narmer to explain the function of a baboon on the middle panel in an underground series below the step pyramid of Djoser. There the king is shown carrying a flail in his right hand, a symbol of his authority over the land, and in his left, an object that seems to be a legal document authorizing a transfer of goods and property. He runs between two sets of three territorial stone markers, representing the limits of the land. By circuiting the markers during the Sed-festival, which took place every thirty years, the king symbolically reclaimed his authority over the furthest boundaries of Egypt. A column of glyphs facing the kings can be translated as 'the White Shrine of the Great Ones.' The shrine seems to have been related to religious and ceremonial functions of the king, particularly in connection with the Sed-festival (Friedman 1995: 24). The identification of the inhabitants of this shrine, the Great Ones, are determined by the image of a squatting baboon, which may represent the body of ancestors that hand over rule to the king. The idea is reinforced by the presence of the legal document carried by Djoser (**Fig. 12**; Friedman 1995: 25-6). Hedjwer, then, may have represented his deified ancestors.

Additionally, the remains of a variety of animals, including baboons, have been revealed at the Royal Cemetery at Hierakonpolis (HK-6), which dates from the Naqada III period. Since Hedjwer merged with the lunar god, Thoth, sometime in the Old Kingdom, these were long believed to be the remains of hamadryas baboons, the species with which Thoth was inextricably linked. However, recent examinations of the skulls from tomb 12 and those found near tomb 2 have revealed that they are actually anubis baboons. This species may have had a range closer to the Upper Nile Valley during the predynastic period, and it may have been easier to obtain than the hamadryas (Linseele 2003). It is possible that at least some of the figurines also portray this species. The word *hedj*, however, means white or silver and is strongly connected with the moon. If the adjective does, in fact, refer to the colour of the god, some figurines may portray the hamadryas because the

[1] There has always been some controversy over the dating system used to discuss the events and rulers of Pharaonic Egypt. For the purposes of this study, the 3rd Dynasty, c. 2686 BCE, will be considered the beginning of the Old Kingdom, while the 20th Dynasty, c. 1070 BCE, will mark the end of the New Kingdom.

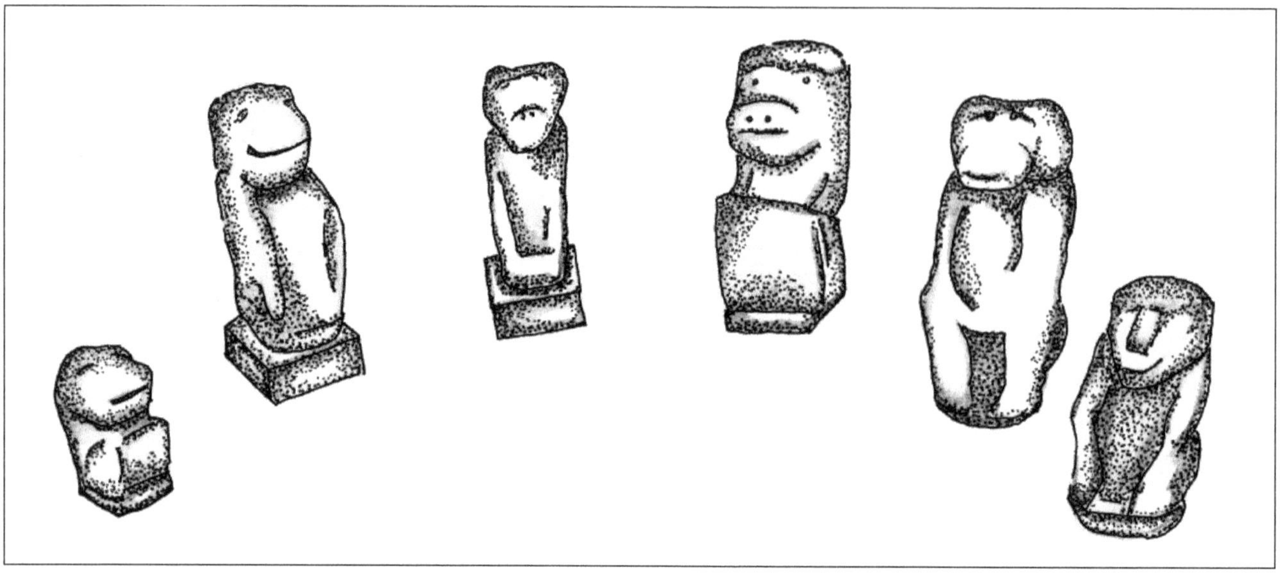

Fig. 9: Glazed composition baboons from the Temple of Osiris at Abydos, c. 3100 BCE (British Museum)

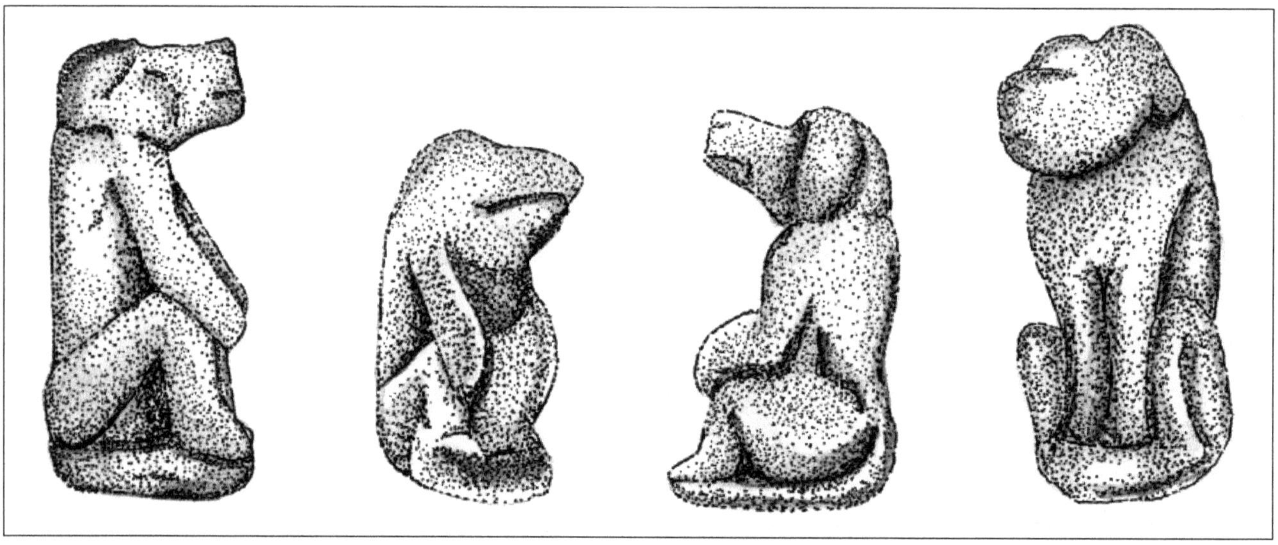

Fig. 10: Glazed composition baboons from Hierakonpolis, c. 2800 BCE (Petrie Museum)

male's fur turns a silvery white as he matures. A few of the figurines, such as UC 11004 (**Fig. 10**), actually seem to have fur covering their ears, which would also indicate that a male hamadryas is depicted.

Stylistically, it is important to note that these early figurines most commonly portrayed baboons in two slightly different seated positions. The animals crouch with their knees bent, and their arms either rest on the tops of their knees or are awkwardly squeezed between their legs so that their fingertips are in line with their toes. Occasionally, the baboons may be holding an object between their knees. The tail is generally curved around the bottom of the figurine. Goudsmit and Scheurleer suggest that the position of the baboons' arms leads to the identification of the species. Because Thoth is later depicted as a hamadryas baboon resting his hands on his knees, they maintain that this is the species depicted in the first type of early figurine. The others, they believe, depict anubis baboons, and for this reason, they suggest that Hedjwer was also an anubis baboon (2000: 24). However, a number of discrepancies in other media make this argument seem improbable. The alabaster baboon with the inscription of Narmer's name, for example, sits with its arms squeezed between its legs but seems to have fur covering its ears—an attribute that characterizes the hamadryas. Secondly, on the relief panel in Djoser's pyramid, the baboon sits with an arm hovering above a bowl, rather than between its legs or over its knees, and its ears seem to be visible. Finally, in later periods, both species of baboon were mummified with their arms resting on their knees. The position of the arms, then, must indicate something other than species.

Recent excavations of the temple at Tell Ibrahim Awad, in the eastern Delta, add even greater complexity to the significance of the baboon figurines. There, approximately two hundred baboon figurines were discovered, but the most remarkable object among them was a glazed-

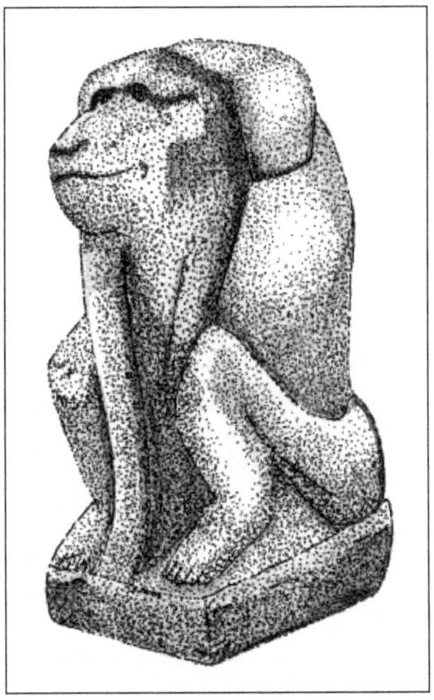

Fig. 11: Alabaster baboon bearing the name of Narmer in Berlin (After photo in Arnst *et al.* 1991: 12, no. 8)

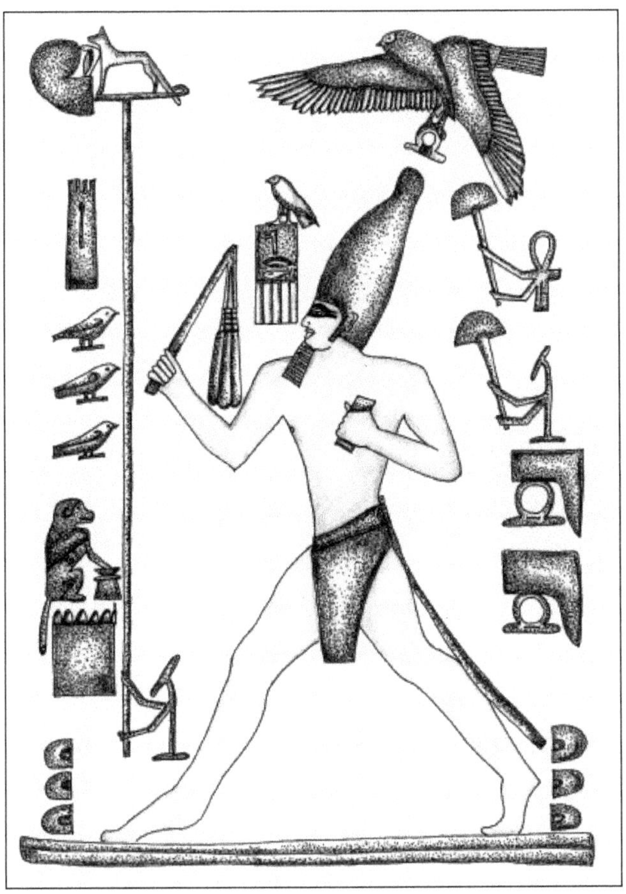

Fig. 12: Middle panel under the pyramid of Djoser (After Friedman 1995: 23, fig. 14)

composition boat containing four baboons. Additional fragments suggest that it originally held seven. Sherkova argues that this boat, which probably dates from the first two dynasties, is early evidence of the baboon's role in the myth of the solar bark's journey through the netherworld. In evidence, she cites spell 126 from the Book of the Dead and an accompanying vignette, depicting four baboons sitting around the Lake of Fire (2000: 504-5). Although I cannot completely agree with Sherkova's argument—she goes on to find a connection between the figurines and Thoth, who was not portrayed as a baboon until much later—it does seem likely that baboons had a solar connection. It has been suggested that the intentional burial of the baboons on the eastern periphery of the cemetery at Hierakonpolis may have reflected an early solar connection (Houlihan 1996: 4-5). Utterance 363 from the Old Kingdom Pyramid Texts mentions the *bnti*, baboon, as the eldest son of Re. This solar baboon seems to be an early manifestation of the *bau* of the east, who greeted the rising sun. Because the Pyramid Texts are more strongly linked to the early dynastic period than the New Kingdom Book of the Dead, this connection seems more probable than Sherkova's attempt to link the baboons to Thoth. More than one god seems to have been associated with baboons at this time, so it seems possible to me that some of the baboon figurines depict hamadryas baboons in association with Hedjwer while others depict anubis baboons, which may have had a solar connection. Another possibility is that the species were used interchangeably during these early times.

Only two examples are known of early figurines portraying standing baboons. The standing figurines were found at Elephantine and 'En Besor, in modern-day Palestine, and only the lower portion of the figurine from 'En Besor is preserved. In both cases, however, the baboons lean toward a cylindrical object, and their tails hang down behind them in a straight line. The Elephantine baboon leans on a receptacle with its left hand and raises its right hand to its mouth (Gophna 1993: 29-31). Although these early artists were not as capable of creating realistic depictions of the animals as later Egyptian artists, the poses could have been taken from nature. It would not be uncommon to see real baboons sitting or standing in these positions.

After these early dynasties, three species of monkeys were commonly depicted, namely hamadryas baboons, anubis baboons and vervets. A fourth, the patas, the only species likely to have inhabited the Upper Nile Valley at this time, also seems to have been represented in tomb chapel scenes. It seems probable to me that this species, which was likely to have been the most commonly spotted monkey in Egypt, was given the least amount of exposure in art because it was the least exotic. The others were imported from the Southern lands of Nubia and Punt, the exact location of which is still unknown. The various species are usually distinguishable in Egyptian art, and baboons are generally assigned roles different from those of smaller monkeys. This chapter, then, will be divided into three main sections: the Old Kingdom and First Intermediate period, the Middle Kingdom and Second Intermediate period, and the New Kingdom. Each section will be further divided into two sections, the first covering non-religious or popular functions of simians in art and culture and the second the hamadryas' role as a manifestation of various gods. In some cases these categories will overlap. Apotropaic amulets, for example, could fit into either category, and some gods revered by the common people were not considered part of the official cult. The final section of this chapter briefly explores developments from the Third Intermediate Period to the Ptolemaic Period.

THE OLD KINGDOM: NON-RELIGIOUS ROLES

Smaller monkeys and maneless baboons (either anubis or female hamadryas baboons) do not seem to have been associated with any deity by the Old Kingdom. They did appear, however, in the form of glazed composition and steatite amulets. Andrews asserts that those depicting a squatting monkey holding a baby are believed to have been designed for female wearers, and amulets of walking monkeys from the First Intermediate Period were worn for their sexual connotations (1994: 67-7). Although she does not explain this interpretation, the monkey's direct connection to women and sexuality is evident in a variety of other media in later periods. Additionally, it is worth remarking that a female monkey's genitalia are most noticeable when she is walking. Baboons, in particular, develop enormous swellings when they are in oestrus and bleed for a few days during menstruation. It would be difficult for anyone who has had frequent contact with monkeys not to notice these sexual characteristics.

Indirectly, monkeys may be related to women through their portrayals on or as containers. As many as fifteen pithemorphic vessels have survived from the Old Kingdom, most of which are in the form of a female vervet clasping a baby to her breast. The majority of these are made from alabaster, and a few bear the names of the Sixth Dynasty monarchs. The cartouche of Pepy II is carved on the mother's right arm on both terracotta and alabaster vessels from Byblos in Lebanon and Balat, while limestone and alabaster examples from Elephantine display the name of Pepy I. Merenre's cartouche appears on yet another alabaster vessel from Qena (Fischer 1993: 1-3). Two fragments of an alabaster vessel were even found in the Peloponnese, at Mycenae. The surviving fragment of the monkey's head is clearly identifiable as a vervet from the exaggerated depiction of the cheek fur. Most of the vessel's front is missing, but an infant's hand is visible on the remaining side (**Fig. 13**; Fischer 1993: 3; Karetsou 2000: 253-4). Many of the mother monkeys are depicted wearing the bracelets, anklets and collars of domestic pets. Often these vessels were used as perfume containers—an appropriate use, as Fischer notes, because both myrrh and vervets were imported from Punt (1993: 9).

An unusual pithemorphic vessel, dating from the Sixth Dynasty, was found in the necropolis at Qila 'al-Dabba and is thought to have belonged to the wife of Ima-Pepi (Valloggia 1993: 396). This is a spouted alabaster cup carved with a vervet clinging to the exterior. The top of the monkey's head peaks over the rim, while the rest of the body can only be seen by looking at the underside of the cup. The monkey squats with its legs splayed on either side of the cup and its arms stretched out to touch the knees. The tail falls in a straight line along the spout, and like most of the monkey jars from the same period, this little vervet is also adorned with bracelets and anklets. An inscription on the monkey's right arm reads, "The King of Upper and Lower, Egypt, Neferkare,[2] living eternally" (**Fig. 14**; Valloggia 1993: 398). Fischer remarks that Old

[2] This is the throne name of Pepi II.

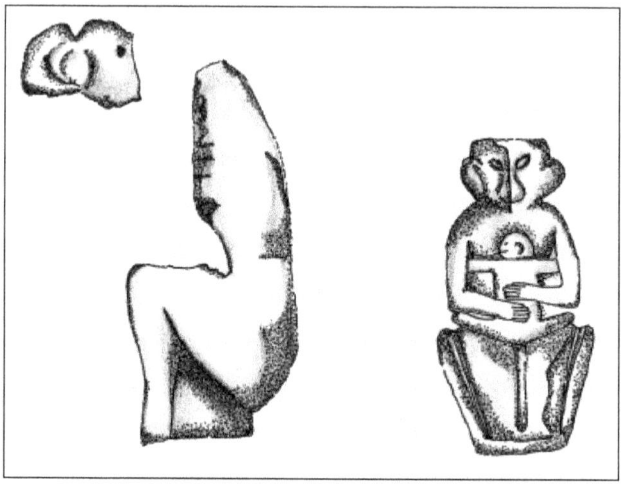

Fig. 13: Fragment of an Old Kingdom pithemorphic vessel found at Mycenae. National Museum, Athens, 2657, 6250 (After Karetsou 2000: no. 252)

Kingdom sculptors avoided placing names on the arms and torsos of human figures—a practice that was common both in the early dynastic period and the Middle Kingdom. He suggests that the names carved on the monkeys simply identified the owner, either the king or someone who had received the vessel as a gift from the king on a festive occasion, such as the jubilee (1993: 7-8).

The most comical and endearing evidence of the roles monkeys played in Egyptian daily life comes from tomb chapel scenes beginning in the Fourth Dynasty. All four species of monkeys are portrayed in these scenes, and all four were clearly kept as pets. Vandier d'Abbadie's series of articles written in the 1960s, "Les singes familiers dans l'Ancienne Égypte," (1964: 147-177; 1965: 177-188; 1966: 143-201) is still the best source of information for these scenes, which have been categorized according to type and period. She established nine categories for the Old Kingdom: monkeys and their keepers, monkeys and dogs with their keepers, monkeys and dogs following the hunt with their master's porters, dwarf keepers, whimsical scenes, monkeys under the master's chair, diverse activities, monkeys in trees, and monkeys on boats (1964: 147-177). Since many of these categories are similar and generally comprise numerous examples, it is not necessary to consider each one individually. In the case of 'monkeys in trees,' only one fragmentary example is extant from the Old Kingdom, and it is very different from Middle and New Kingdom examples of the same type, which are considered in their appropriate sections.

The first three involve servants holding monkeys by a leash or simply accompanying the animals. Since several texts mention monkey herders or guardians, it is likely that these servants' only duty was the care of the animals. They should be differentiated from the trusted servants leading monkeys, who are depicted in other Old Kingdom scenes following their masters with a sack containing his fresh change of clothes. The ordinary monkey herders were apparently notorious for sexual assault (Vandier d'Abbadie 1964: 152). One of the most hilarious examples

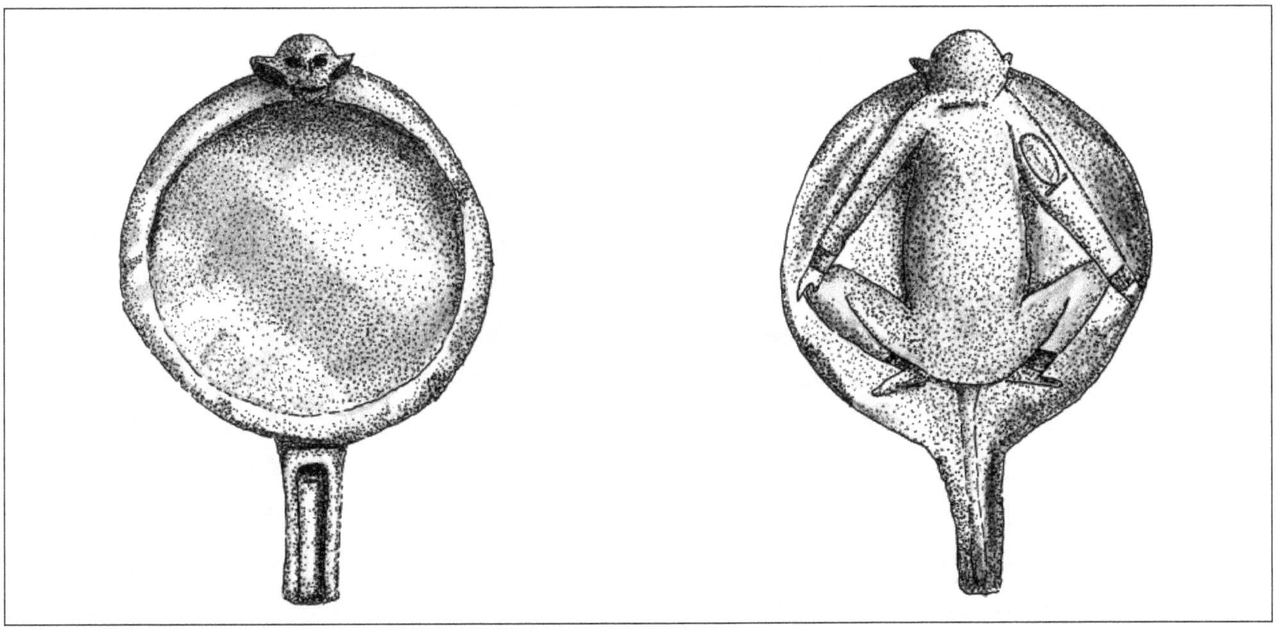

Fig. 14: Spouted cup from Balat (After Valloggia 1993: 398)

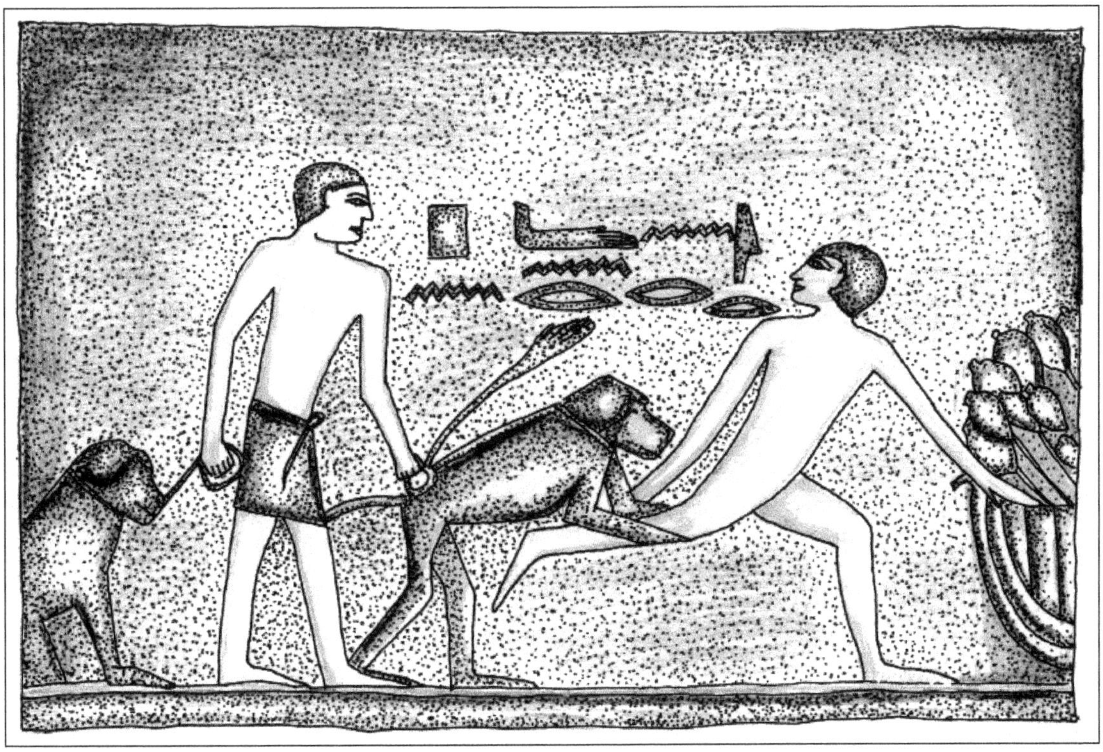

Fig. 15: A baboon catches a thief at the market (After Houlihan 2001: fig. 6)

of these scenes comes from the Fifth Dynasty mastaba of Tepemankh at Saqqara. In this scene, a man equipped with a baton terminating in the shape of a human hand leads a pair of anubis baboons attached to leashes through a marketplace. The male grasps the leg of a nude fleeing thief, who tries to wrench the baboon's arm from his leg. The female remains behind her keeper and clasps an infant to her chest (**Fig. 15**). The nature of this scene is somewhat controversial. Some scholars maintain that this is a scene depicting a policeman with his baboon assistants (Janssen and Janssen 1989: 25), but Houlihan asserts that the man is probably a market attendant exercising his pet baboons and that the scene adds a touch of humour to an otherwise formal composition (1996: 106).

Dwarf guardians are an interesting variation on the normal-sized handlers and are depicted in the same scenes, carrying the same type of baton. Although scenes such as

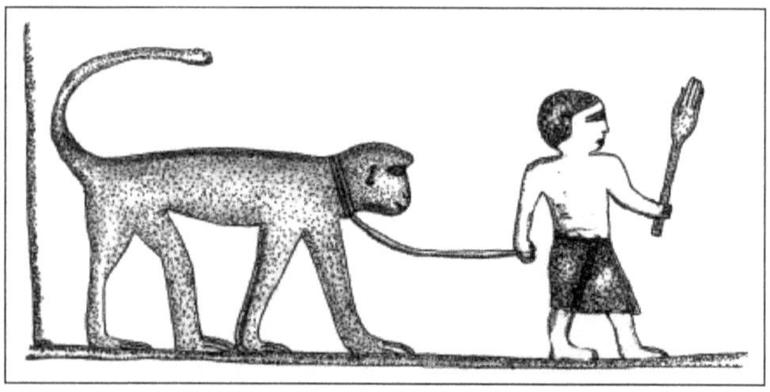

Fig. 16: A dwarf monkey keeper from the tomb-chapel of Ti (After Houlihan 1996: pl. 6)

the painted relief from the Fifth Dynasty tomb chapel of Ti portray the monkey as twice the size of the dwarf handler, it is clear from the morphology of the animal that one of the smaller species of monkey is represented (**Fig. 16**). In fact, dwarf keepers are not portrayed with baboons at all (Dasen 1993: 116). Perhaps these large animals would have been too powerful for them.

A scene from the tomb of Pepyankh at Meir is an excellent example of light-hearted whimsy. This could also belong to the third scene type, since it is a case of monkeys and dogs following the hunt with their master's porters, and both ordinary monkey keepers and the trustworthy servants carrying their master's laundry sacks are shown. A small monkey, whose leash is held by a keeper, rides on top of the palanquin, while another sits on the shoulders of the sack carrier. In another scene from Meir, a monkey stands on the back of a dog to tweak his keeper's nose (**Fig. 17**; Vandier d'Abbadie 1964: 163-4, figs. 19, 20). In other humorous scenes, dwarves may be present. The Sixth Dynasty tomb of Ankh-ma-Hor depicts a vervet sitting on the shoulders of his dwarf handler, eating a fig that he may have stolen from the handler's basket (1964: 165, fig. 22).

The favourite spot of vervets appears to be under the master's chair. Numerous scenes depict the pampered creatures sporting collars, and occasionally bracelets and anklets, munching away on fruit placed in baskets before them, while their master sits above. Sometimes they are joined by their dwarf handlers, as in the scene from the mastaba of Idu, in which a monkey is perched on the head of a dwarf, while Idu sits on a chair above them in the company of his wife (**Fig. 18**; Vandier d'Abbadie 1964: 166, fig. 24). In some cases, the monkey is actually under the seat of the tomb owner's wife, which leads Manniche to suggest that the monkey may have an indirect erotic implication in such scenes. She argues that monkeys were often associated with the dwarf god, Bes, who protected women in many aspects of their private lives (1997: 43-4). However, she does not make a distinction between similar scenes from different periods, and the connection is slightly anachronistic if applied to the Old Kingdom. Despite the early existence of dwarf gods, Bes is a Middle Kingdom creation, whose association with monkeys does not seem to have been common before the Third Intermediate Period (Dasen 1993: 57; Romano 1989:

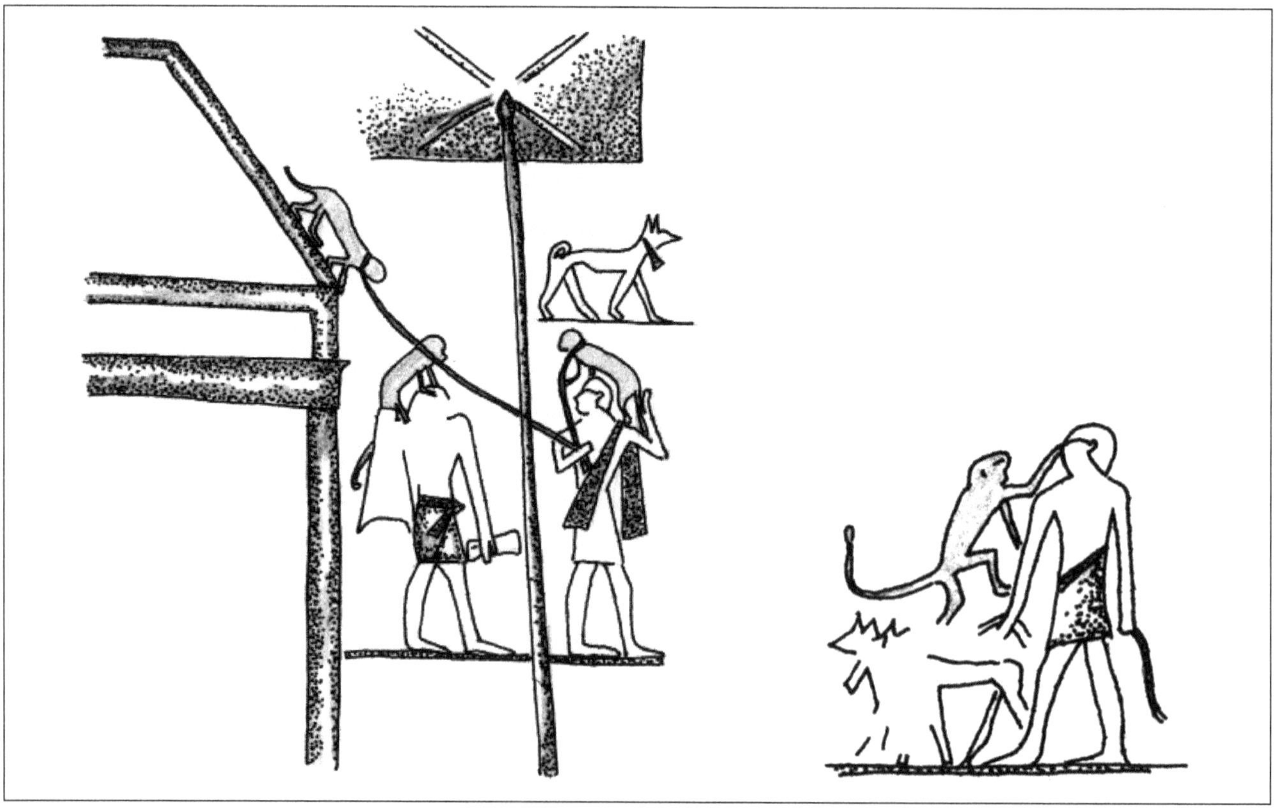

Fig. 17: Scenes with monkeys and their keepers from the tombs of Meir (After Vandier d'Abbadie 1964: 163-64, figs. 19 & 20)

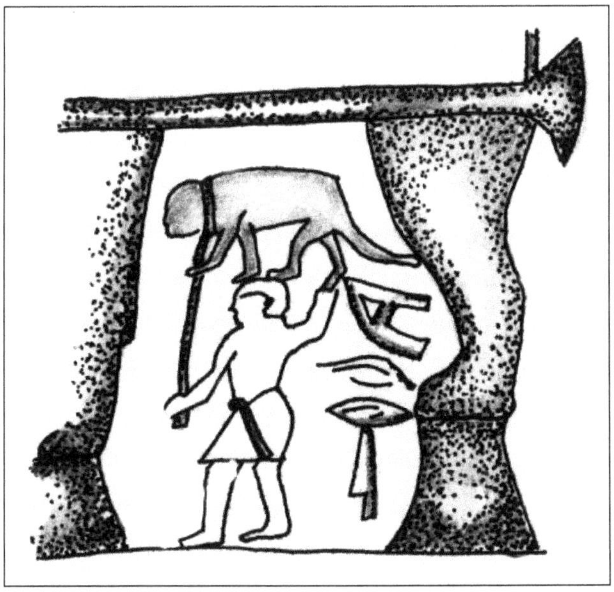

Fig. 18: A monkey and dwarf under the chair of Idou (After Vandier d'Abbadie 1964: 166, fig. 24)

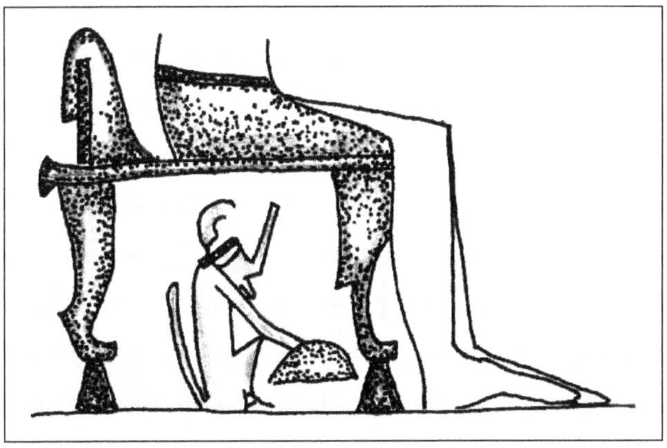

Fig. 19: A female monkey under her master's chair (After Vandier d'Abbadie 1964: 167, fig. 26)

133). Therefore, while monkeys did have some connection to women through amulets and vessels at this time, it is difficult to argue their erotic value in Old Kingdom tomb chapel scenes on the basis of their appearance with Bes on amulets from a much later period. One scene, however, does seem to emphasize the female characteristics of a pet vervet. In this Sixth Dynasty scene, the monkey, who sits beneath her master's chair eating fruit from a basket, is portrayed with a pendulous breast. She wears jewellery, a necklace rather than a collar, and bracelets and anklets (**Fig. 19**; Fischer 1993: 3-4, fig. 3; Vandier d'Abbadie 1964: 167, fig. 26).

The next category comprises monkeys engaging in diverse human activities for the sake of entertainment. For the most part, these are scenes of music and dance. During this period, the only instrument monkeys are actually shown playing is the flute, and these images are found only on Fifth Dynasty cylinder seals (Vandier d'Abbadie 1964: 169). Nevertheless, simians are shown imitating the moves of dancing girls and even mimicking the gestures of a dwarf conductor (1964: 169-70). A scene from the Sixth Dynasty tomb chapel of Niankhpepy depicts a baboon walking on two legs and carrying a yoke with bundles at either end. While Vandier d'Abbadie suggests that baboons might have been trained to carry burdens (1964: 170), I am inclined to agree with Houlihan, who argues that such scenes should be considered examples of humour (2001: 20-1). While baboons could be trained to perform such an act on an occasional basis, the animals are not comfortable walking upright for long periods of time.

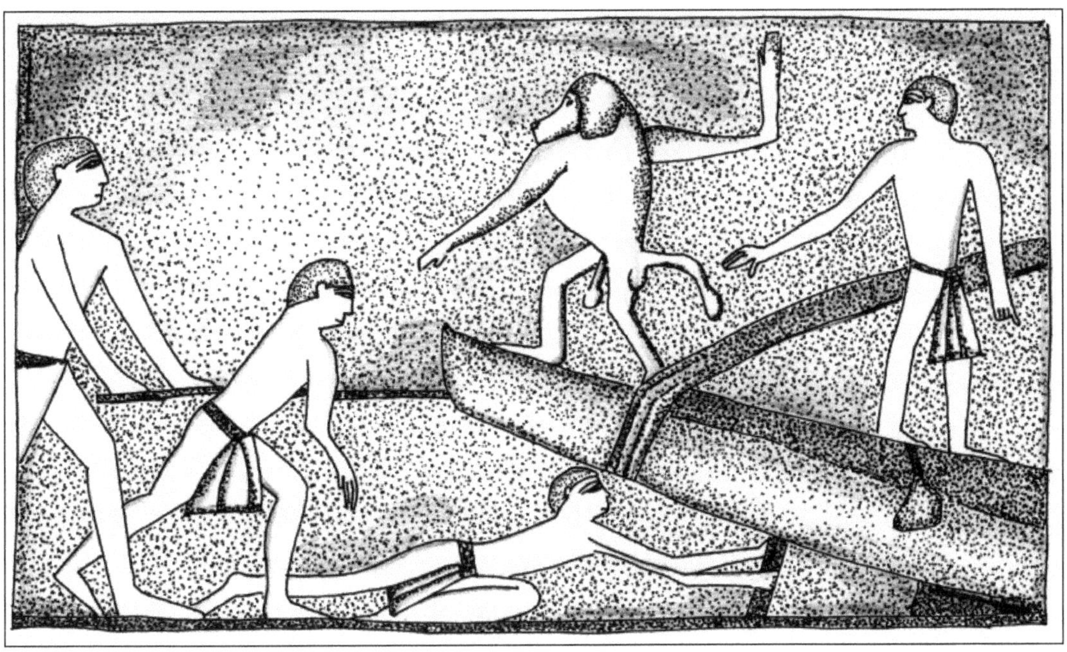

Fig. 20: A baboon acts as an overseer in a ship-building scene in the tomb-chapel of Nefer (After Houlihan 2001: 19, fig. 7)

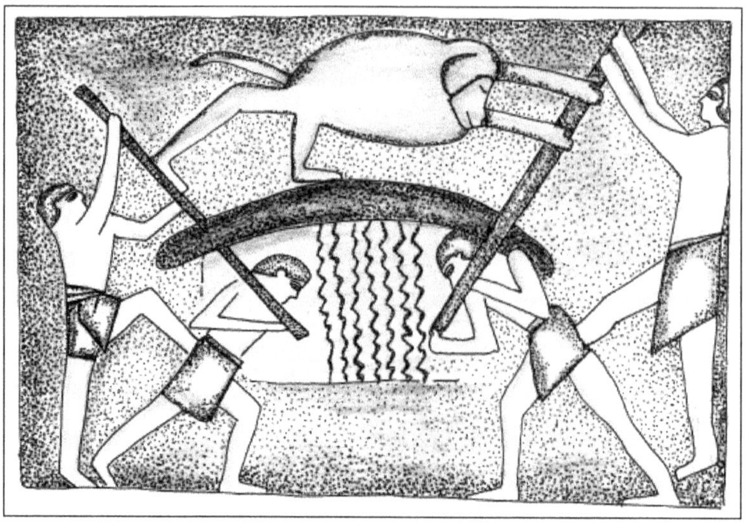

Fig. 21: A baboon helps twist a must-sack in the tomb-chapel of Nefer (After Houlihan 2001: 20, fig. 8)

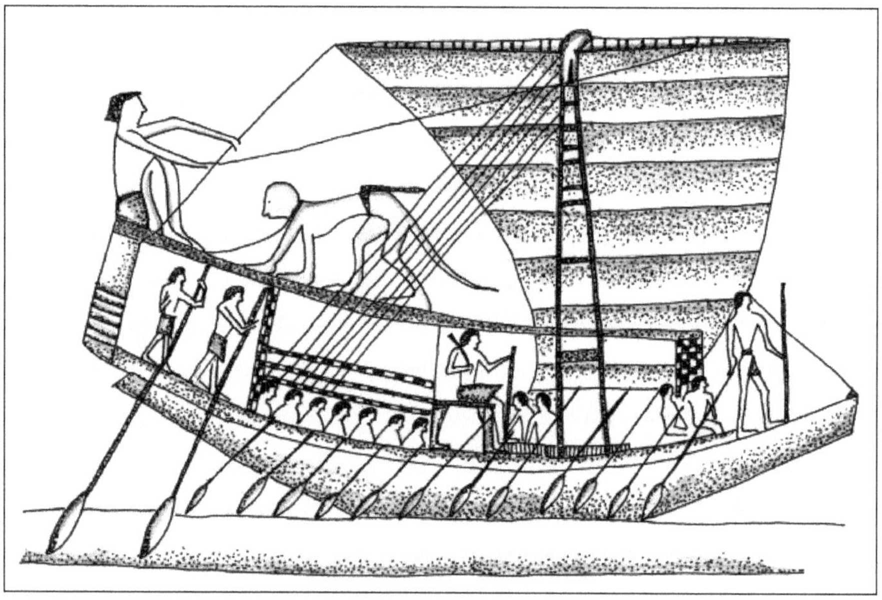

Fig. 22: A giant vervet walks on a boat in the tomb of Snefru-ani-mertef (After Vandier d'Abbadie 1964: 174, fig. 39)

Scenes in limestone relief from the Fifth Dynasty tomb chapel of Nefer at Saqqara more unequivocally depict simian parodies of human behaviour. In one scene, a large male hamadryas baboon takes on the role of an overseer in a ship-building scene. He stands fiercely on the stern of a boat with a large baton raised over his head, apparently directing and threatening the human workmen nearby (**Fig. 20**). A second scene from the same tomb chapel depicts the hamadryas in a more friendly light. A group of four men are laboriously pulling poles attached to a must sack to extract the grape juice. The gigantic hamadryas stands on top of the sack with one foot and presses one of the poles with his free leg, while forcing the other pole backward with both hands (**Fig. 21**; Houlihan 2001: 19-29, figs. 7, 8). His benign expression is in striking contrast with that of the baboon overseer. It may not be possible to know whether the same baboon is depicted in both scenes, but the artist cleverly captured something of the dual nature of baboons. While the scenes are clearly caricatures, the baboon's strength is something that can be both intimidating and, on occasions, helpful.[3] The baboon's virility is also apparent in both scenes from the huge but realistically rendered male organs.

The next category, monkeys on boats, depicts vessels coming from the south with their cargo of live vervets. The animals are shown in perfect freedom on the boats, climbing the ropes and sitting on the masts. In the Fourth Dynasty scene from the tomb of Snefru-ani-mertef, a vervet walks across the top of the vessel. Vandier d'Abbadie notes that if the monkey were drawn to scale, it would certainly cause the boat to capsize and remarks that it is another example of the Egyptian indifference to scale in the rendition of their subjects (**Fig. 22**; 1964: 174, fig.

[3] Whenever we built indoor cages at our sanctuary, we would test the strength of a frame by letting our baboons play on it for a while. Often, much to the dismay our hired helpers, the baboons would demolish in half an hour a structure which had taken hours to erect.

39). It seems unlikely, however, that the massive vervet is simply the product of indifference. In Egyptian art, differences of scale are generally not reflections of reality. Gods and kings are depicted as larger than their subjects, and when there is an enlargement of a certain object within a scene, this may reflect its importance in the mind of the artist (Wilkinson, R. 1999: 38-9). Such is most likely the case with this scene. Although it is impossible to know exactly why the vervets are portrayed on a massive scale, it seems probable that it is their value as an exotic pet and symbol of wealth that is being portrayed here.

Lastly, mention should be made of the single Old Kingdom example of a monkey in a tree. This fragmentary scene from the Fourth Dynasty tomb of Nefermaat depicts a man gutting an animal that is hanging upside down on a tree branch. A vervet can be seen climbing the other side of the tree (**Fig. 23**). Vandier d'Abbadie remarks that, although the tree appears to be a sycamore fig tree, the monkey does not seem to be gathering fruit (1964: 171, fig. 35). This seems to have been a common activity for monkeys, especially baboons, in later periods.

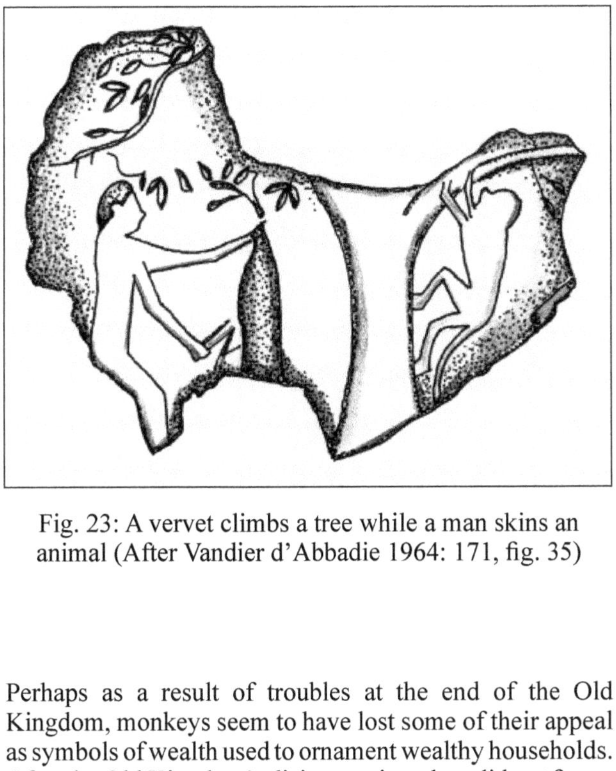

Fig. 23: A vervet climbs a tree while a man skins an animal (After Vandier d'Abbadie 1964: 171, fig. 35)

Perhaps as a result of troubles at the end of the Old Kingdom, monkeys seem to have lost some of their appeal as symbols of wealth used to ornament wealthy households. After the Old Kingdom's disintegration, they did not figure in tomb scenes again until the Twelfth Dynasty and Second Intermediate period, where scenes are most prevalent in the tombs of nomarchs at Beni-Hasan and public officials at el-Kab (Vandier d'Abbadie 1965: 12).

THE OLD KINGDOM: BABOONS AS MANIFESTATION OF DEITIES

The Pyramid Texts makes reference to a number of baboon deities at this time. The *bnti* (baboon) is named as a son of Re in Utterance 363, and in Utterance 569, two deified baboons, *bnti.wi*, again sons of Re, are mentioned in an exhortation to Re to allow the coming of the deceased king (Mercer, Trans. 1952):

> The (coming of) the two apes (*bnti.wi*) to Re, his two beloved sons, shall be prevented if thou preventest N. From coming to the place where thou art.

These are perhaps the same "apes" who sit in the east of the sky to adore the sun at its rising. In Utterance 315, they are addressed by the deceased Pharaoh (See Faulkner's trans., 1998).

Although baboons were considered in a positive light in most cases, their negative aspects were also acknowledged—the baboon in hieroglyphic script could be used as a determinative for aggression (Jannsen and Jannsen 1989: 21). Furthermore, the modern word 'baboon' may have come from the name of an aggressive deity called Baba (Wilkinson 2000: 73), alternatively known as Babwi, or Babi, whose original manifestation in the Pyramid Texts was that of a baboon. Utterance 549 of the Pyramid Texts reproaches his thievery (Faulkner, Trans. 1998):

> Get back, Babi, red of ear and purple of hindquarters! You have taken the thigh joint of your goddess to your mouth!

Derchain remarks that it may be Babi's red ears that set him apart from other baboons (1952: 27), but I suspect the colour of the ears and hindquarters (red and purple are the same word) served to identify the species of baboon because the hamadryas does have pink ears.[4] The colour may, instead, serve to emphasize the sexual potency and aggression of the male baboon. Red, in Egyptian art, had a number of connotations. It could represent fire, blood, life, and regeneration, but it could also symbolize the harsh desert areas, anger, and destruction. Red ink was used to write the hieroglyph for "evil," and the colour was also associated with other hostile deities, such as Seth and Apophis (Wilkinson, R. 1999: 106), and in the New Kingdom, phallic votive objects were sometimes painted red (Meskell and Joyce 2003: 110).[5] Although the colour of Babi's phallus is not mentioned in the Pyramid Texts, its function is described as the bolt that locks the doors of the sky (Faulkner, Trans. 1998: Utt. 323). Meskell groups Babi with a number of other phallic deities, such as Min and Seth, who represent sexual power rather than fertility (Meskell and Joyce 2003: 105). Babi's strength and status, emphasized in Utterance 320, set him apart from other baboons (Faulkner, Trans. 1998):

> ....for the King is Babi, Lord of the night sky, Bull of the baboons who lives on those who do not know him.

Clearly, Babi is the chief baboon, the most powerful. The King attempts to assume his power by declaring that he

[4] This, of course, is more obvious in young baboons that have not yet developed the heavy mantle that covers the ears.

[5] It is interesting to note that baboons themselves consider red to be an impressive colour. Kummer noticed that geladas coexisting in the wild with the hamadryas would give way to the latter, even when they were more numerous. He explains that male geladas have a bare, pink area on their chests that acts as a stimulus to their conspecifics. Their hindquarters and faces, however, are dark brown, while those of the hamadryas are bright red. The colour is emphasized by the striking white hair surrounding the face. He suggests, therefore, that the geladas perceive these brightly coloured animals as "supermales" (1995: 67).

is Babi, a fierce god who threatens to consume those who do not know him. As time progressed, Babi's behaviour became more malevolent. By the New Kingdom, he developed an association with Seth and adopted a dog-headed manifestation. He became the adversary of a number of benign deities, especially Thoth, which will be discussed in the following sections.

As mentioned above, it is generally accepted that Thoth began to merge with Hedjwer in the Old Kingdom. However, his association with baboons was not particularly apparent at this time, and because Hedjwer continued to maintain a distinct personality even in a few spells from the Middle Kingdom Coffin Texts, it is difficult to guess which aspects of the baboon god Thoth acquired at this time. Throughout Egyptian history, Thoth was most commonly depicted in his anthropomorphic, ibis-headed form holding a scribal palette and pen or notched palm leaf, in the act of calculating or recording. The ibis' long, curved beak may have been associated with the crescent moon and reed pen (Bleeker 1973: 110). Later, the baboon was also used as a visual metaphor for Thoth to represent qualities associated with him. In the hieroglyphic script and in representational works, the baboon could signify wisdom, knowledge, judgment, writing or excellence (Wilkinson, R. 1999: 157). His name, however, is usually written in hieroglyphics with the form of an ibis and read as Djehuty (Bleeker 1973: 106). In order to understand how Thoth's baboon association began, it may be helpful to explore some of the evidence concerning the god's origins.

Bleeker notes that the ibis appears in the coat of arms of the 15th nome of Lower Egypt. The capital, later called Hermopolis Parva, may have been his original cult centre (1973: 109). It is impossible to know whether Thoth was always associated with the ibis or whether he took over the role of a local ibis god. In addition to the bird's lunar associations, Bleeker remarks that the ibis was venerated because, according to Plutarch (*De Iside et Osiride* 75), it attacked poisonous reptiles and never drank impure water (Bleeker 1973: 110). Like the ibis, Thoth was always acknowledged to be an opponent of hostile forces who also possessed exceptional wisdom.

The god may have begun to acquire his baboon association as early as the Old Kingdom when Hermopolis Magna became his main cult centre, since this had been Hedjwer's centre of cult throughout the early dynastic period. It is unclear, however, why Thoth later became so strongly connected with the baboon. Boylan suggests the animals were symbolic of Thoth's mysterious knowledge (1922: 97). Bleeker disagrees, revealing personal bias against the species, "It is doubtful whether the ape shows any evidence of exceptional wisdom" (1973: 111).

Although Thoth is not associated with baboons at all in the Pyramid Texts, his many functions and attributes are still worth considering, since, by the New Kingdom, his baboon manifestation had acquired many of the traditions that had once been associated with the ibis. Even at this early period, Thoth had important functions in the afterlife. As an ibis, he played the role of psychopompos for the deceased king, who was carried on his wings (Pyramid Texts, Utts. 270, 555, 556).

His roles in the myths of Osiris and Horus are, of course, the best known. In Utterance 477, Horus and Thoth raise Osiris from his side (a euphemism for death), and cause him to stand as chief among the Two Enneads. The myth of Thoth's rescue and restoration of Horus' eye after the latter's battle with his uncle, Seth, is related in Utterance 359. In these traditions, Thoth acts not only as an assistant, but as a magician who can raise the dead, a healer who can restore the wounded eye, an advocate for the righteous, peace-maker and arbitrator.

Nevertheless, it would be misleading to suggest that Thoth was always considered a beneficent deity. Utterance 218 suggests an alternate myth in which Thoth is linked to Seth as an enemy of Osiris (Mercer, Trans. 1952):

> Behold what Set and Thot have done, thy two brothers, who knew not how to weep for thee.

This version, though, does not seem to persist after the collapse of the Old Kingdom, and only the positive attributes mentioned would become linked with Thoth as a hamadryas.

THE MIDDLE KINGDOM: POPULAR ROLES

By the Twelfth Dynasty, a certain amount of democratisation was taking place in society, and new stone working techniques were becoming available to artists outside the boundaries of the palace. Bourriau remarks, "Objects made in the Middle Kingdom differ most strikingly from those produced at earlier or later times by reason of the range and quality of materials available to the craftsman" (1988: 127). With the acquisition of semi-precious stones from the eastern and western deserts, Middle Kingdom artists were able to create stunning pieces of jewellery and household items, such as anhydrite kohl pots. While some of the jewellery was clearly created in workshops attached to the palace, some found in the burials of private individuals suggests that fine jewellery was also produced away from the capital in locations such as Beni-Hasan. At this time, more than at any other, private individuals were buried with their jewellery (Bourriau 1988: 127-8).

Naturally, these changes in society and material had a noticeable effect on the representations of monkeys. The vervet and maneless baboon amulets that had been produced in glazed composition and steatite during the Old Kingdom began to be fashioned from amethyst and cornelian (Andrews 1994: 66). Faience, however, did not disappear. Apotropaic animal figurines, such as lions and baboons, continued to be hand modelled in faience. Many of these animal statuettes seem to depict the same protective demons that appear on apotropaic objects known as magic knives or wands and may have been a necessary component to rituals associated with them (Bourriau 1988: 116). The wands, produced in the Middle Kingdom, were made from hippopotamus ivory, and are similar in shape to throwsticks used for hunting birds. Some have abrasions on the pointed end, which suggest that they were employed to scratch lines in the earth, probably protective circles. Clay models of the wands were sometimes interred with the dead (Pinch 1994: 78). Their primary function, though, was to protect society's most vulnerable individuals, particularly pregnant and nursing woman and children (Bourriau 1988: 114; Weingarten 1991: 4). Weingarten notes that the quality of the engraved images on the wands is often poor—a surprising fact when one considers that even hippopotamus ivory was considered a luxury item and that the women who owned these objects were often of high rank (Weingarten 1991: 4). The most common demon on the wands was the hippopotamus demon that later became Taweret (1991: 4). This creature had the head and body of a hippopotamus but walked upright on her hind legs and carried a knife in one hand. Her mouth is usually open to reveal her formidable teeth, her breasts sag towards her protruding belly, and her back is covered with a long appendage. An understanding of the nature of this demon may be useful because baboons were often carved on the same wands and later depicted with images of Taweret in the New Kingdom. In utterances 381-382 of the Pyramid Texts, a hippopotamus demon called Ipy acted as a nurse to the deceased Pharaoh (Faulkner, Trans. 1969).

Weingarten explains that milk is a protective fluid and that the demon's earliest function was to protect and nourish. By the Middle Kingdom, two coffins depict the knife-wielding hippopotamus speaking the following spell (Faulkner, trans. 1969 cited in Weingarten 1991: 10):

> I am Many-faced who created thunder, who mounts up to Re and repels the strength of 'Apep, who splits open the sky and drives away storm, and who nourishes the crews of Re.

Clearly, then, the demon's presence on the wands carried amazing protective qualities. As was the case with many dangerous animals, the hippopotamus' power and destructiveness were used as weapons against hostile forces. Even the ivory from which the wands were made provided the owner with the power of the hippopotamus (Wilkinson 1999: 92). Perhaps the animal's naturally protruding belly led the Egyptians to think of pregnancy, and the fact that the animals spent most of their time in the water lead to an association with life-giving fluids. The Nile god, Hapy, for example, is also depicted with pendulous breasts.

The baboons seem to have had similar apotropaic functions, probably because the animal's potential ferocity and enormous canines would make it a formidable guardian. A wand from the collection of the Metropolitan Museum of Art depicts a series of weapon-carrying demonic animal divinities, including lions, vultures, the hippopotamus demon, and seated male hamadryas baboons holding braziers (MM 08.200.19 in Weingarten 1991, pl. 5). Another unfinished wand from Thebes, now in the Royal Museum of Scotland, begins with the *sa* hieroglyph of protection, formed from a reed shelter. A hamadryas baboon walks toward it carrying another *sa* sign. Behind him are the Seth animal, another *sa* sign, and the *wedjat* eye (1921.893 in Bourriau 1988: 114, fig. 102). The baboon's apotropaic roles are not as well defined as those of the hippopotamus demon at this time, but the baboon's proximity to the *wedjat* eye may relate to Thoth's journey into the desert to retrieve the solar eye—a story for which there is no written record before the New Kingdom—or his rescue of the lunar eye of Horus (Pinch 1994: 41). The hieroglyphic sign of the brazier, held by the baboons on the Metropolitan Museum wand, could be used as a determinative for different aspects of heat or fire. In addition to solar energy, fire was an important component in the underworld (Wilkinson 2000:161). In later periods, vignettes from the underworld books would depict both braziers and baboons surrounding a lake of fire.

Anhydrite was another material that became available only in the Middle Kingdom. Despite its common name of blue marble, its colour ranges from an almost colourless blue to violet to white tinged with brown or red. In Egypt, flawless blue anhydrite was highly prized, as was its ability to take a high polish. No examples are known before the Eleventh Dynasty, and it seems to have faded from use after the height of its popularity in the Twelfth Dynasty. The stone was used almost exclusively for toilet jars, most of which are unadorned. The few jars that were decorated, however, usually incorporate the shapes of ducks or monkeys (Terrace 1966: 57-62). While a few vessels are actually pithemorphic in the style of the Old Kingdom alabaster jars, most are decorated with smaller monkeys or baboons in relief stretched across the outside

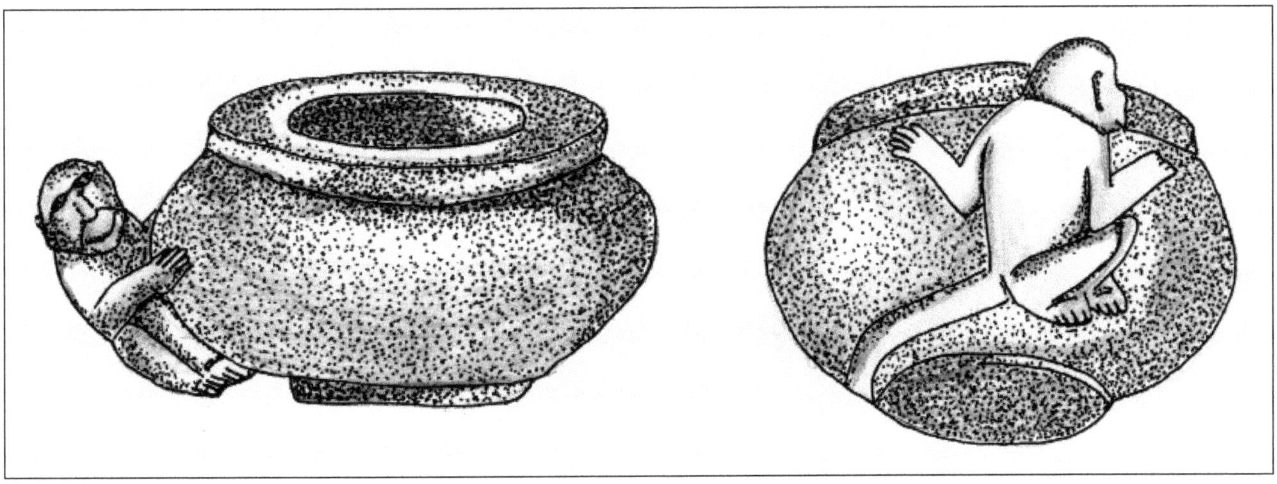

Fig. 24: A blue marble jar with a monkey clinging to the side
(After Terrace 1966: pl. 14, figs. 1 & 2)

of the container (Terrace 1966: 57-63). Often the monkeys depicted on these containers can be identified as baboons. Bourriau describes two similar Twelfth Dynasty kohl pots, which both have two baboons carved in relief around the outside of the container. The animals are seated on their tails with their arms and legs bent. Their fingers and toes are nearly touching as they hug the vessel. The baboons are also wearing collars, suggesting that they are domesticated animals (1988: 142, figs. 144a, 144b). On one unusual jar, a single monkey, carved almost completely in the round, clings to the side only with its arms. It legs are held tightly together and raised under its right arm. Both feet rest on the side of the jar, while its long tail wraps around the base. The monkey seems to have the muzzle of a baboon, but its tail is rather long (**Fig. 24**; Terrace 1966: 59, Pl. 15, fig. 1). It seems possible to me that the baboons on these vessels may also have an apotropaic function. Eye infections were common in Egypt, and these baboons may have served to protect the eye make-up in addition to adding a touch of whimsy to the containers.

Monkeys continued to be associated with perfume. An anhydrite ointment jar in the Metropolitan Museum of Art was carved in the form of a seated baboon holding a jar. It wears a shell pendant that was a badge of office. The animal smiles benignly (**Fig. 25**; 10.176.54, Manniche 1999: 70). Other ointment jars have baboons carved in relief around the outside of the vessel in the same manner as the kohl pots. Some vessels have as many as four baboons (Terrace 1966: Pl. 18, figs. 11-12).

The monkey's associations with vessels, cosmetics and women are also apparent in tomb chapel scenes. A scene from the tomb of Sebeknekht at el-Kab depicts a vervet helping a servant dress her mistress's hair by holding the unguent jar for her (Vandier d'Abbadie 1965: 181, fig 6), and another from the tomb of Bebi depicts a monkey with curiously striped fur, sitting under the seat of Bebi's wife, holding an unguent jar. A mirror is displayed under the seat, and a servant girl is shown standing behind the lady, arranging her hair (Vandier d'Abbadie 1965: 184, fig. 7).

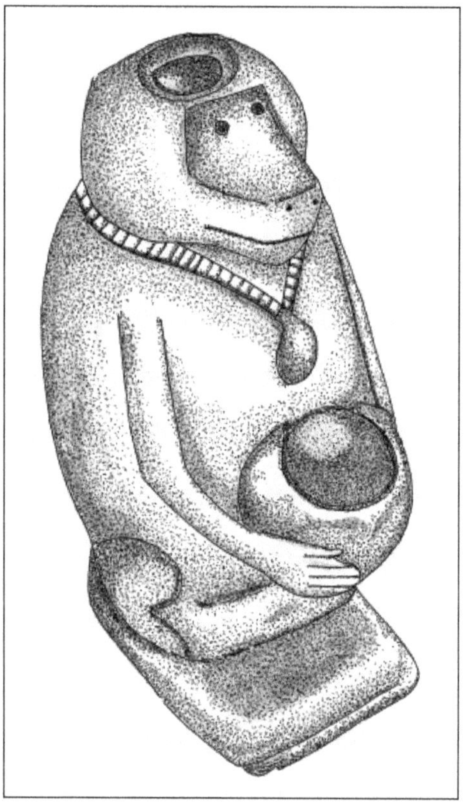

Fig. 25: A blue marble unguent jar in the shape of a baboon (After Manniche 1999: 70)

Many of the themes of the Old Kingdom tomb chapel scenes, however, disappeared at this time. Only scenes of monkeys or monkeys and dogs with their keepers remained relatively popular, and dwarves no longer appeared with the animals. Only one scene is known of a monkey under its master's chair, and scenes of music and dance are also noticeably absent.

One remarkable scene from this period comes from the tomb of Khnumhotep at Beni-Hasan, in which three baboons, probably anubis because of their greenish-brown fur and lack of mane, are shown eating figs in a tree, while two men on either side of the tree collect fruit in baskets. One man is shown standing by the tree and reaching for a fig, while the other kneels by his basket with his hand lightly touching a fig (**Fig. 26**). This scene represents the first example of a theme used extensively in the New Kingdom, the exact nature of which is unclear. Many scholars argue that the monkeys in scenes such as this are helping to pick the fruit and that the baboons were trained specifically for this purpose. The major opposition to this notion comes from Patrick Houlihan, who, in his article, "Harvesters or Monkey Business," summarizes and refutes arguments in favour of monkey harvesters (1997: 31-47). Here I would like to offer a counter refutation to one of his main arguments.

While Houlihan acknowledges that pig-tail macaques (*Macaca nemestrina*), have long been employed as assistants in coconut harvests in parts of Southeast Asia, he claims that they do not assist with the harvest of soft fruits. He also remarks that because the baboons are eating the fruit, they are probably helping themselves but not the men below (1997: 32-37). However, in the past decade, farmers in Thailand have begun training macaques to harvest tamarinds and mangoes. Training takes six months to a year, during which time the animals are rewarded with fruit and affection. To harvest the fruit, macaques must climb trees the height of a three story house and shake the branches. While they are in the trees, they may eat as much fruit as they like (CNN.com, 2000). Although monkeys have only recently been used to pick soft fruits in Thailand, they had been engaged in the activity in other parts of Southeast Asia prior to Houlihan's article (Lindburg 1980: 166). Furthermore, in other discussions of this topic, it has never been remarked that baboons, like macaques, have cheek pouches for storing food. Generally, these animals will only take as much food as they can fit in their pouches and chew on it later while performing other activities. It would be economically sound, therefore, to allow the monkeys to take as much fruit as they wanted.

Before abandoning this scene, it is worth considering one more aspect of the argument. Houlihan states that the kneeling man is not picking up figs that the baboons have thrown to the ground, and, indeed, no figs are shown lying

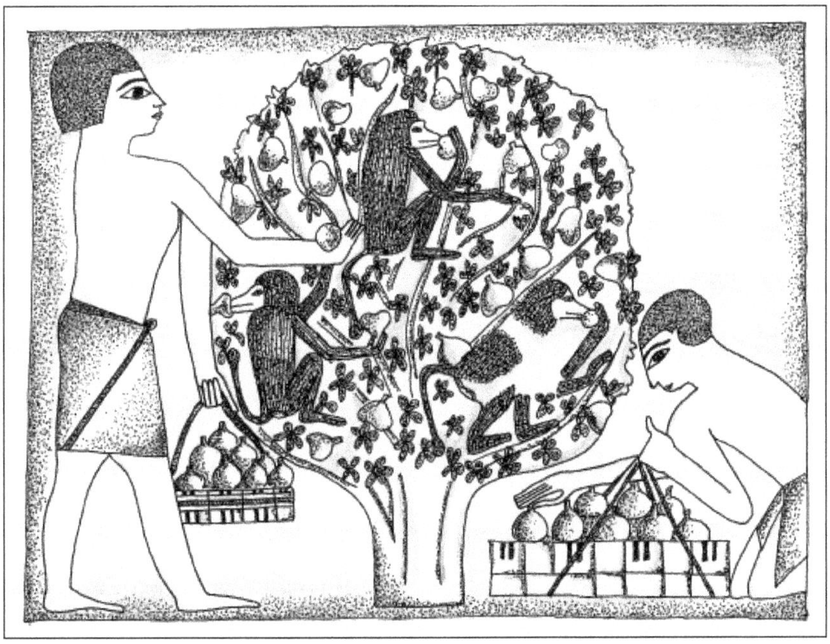

Fig. 26: A scene from the tomb of Khnumhotep at Beni-Hasan

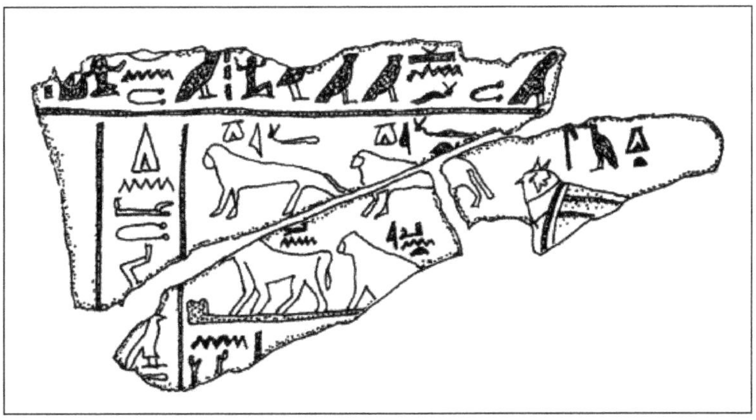

Fig. 27: Scene from the tomb of Nehera (After Griffith and Newberry 1937: pl. 11, fig. 5)

on the ground (1997: 34). However, he fails to remark on other elements in the scene that seem incongruous. One of the most noticeable is the fact that the men, tree and baboons are not drawn to scale. A mature fig tree is much taller than a man, yet the harvester on the left-hand side of the scene is the same height as the tree. This allows him to make the instantly recognizable gesture of reaching out to pluck a fig. The baboons are much smaller in proportion to the men than they would be in real life. Additionally, a small bush could not support three anubis baboons—even females are enormous! It seems more likely that the scene was created to be read differently. Perhaps the three baboons signify that the tree was large enough to support a number of animals—three symbolized plurality—while the size of the men and their baskets reflect the importance of the harvest. The artist may have been disinclined to include the step of retrieving fallen figs because the viewer would have understood the scene without it. I do agree, however, that a touch of gentle humour is present in the scene—the baboons clearly love their work, and all three are smiling!

Lastly, mention should be made of two unusual scenes from the Twelfth Dynasty rock-tombs at el-Bersheh and Meir. The first scene, from the tomb of Nehera, is very fragmentary, but it depicts both vervets and male and female baboons, followed by a mythical *sag* animal (**Fig. 27**; Griffith and Newberry 1937: Pl.11, fig. 5). In the second scene from the tomb of the nomarch, Ukhhotep, at Meir, a monkey is giving birth (**Fig. 28**). This seems to be one of the few examples of a monkey shown in a setting that is neither domestic nor religious. The simian, according to Vandier d'Abbadie, appears at the bottom of a scene, depicting a hunt in the desert. Other animals, such as lions, oryx, and a stag already wounded by an arrow, are attacked by unleashed dogs. She finds the inclusion of the monkey strange, since, she states, monkeys live in forested areas (1965: 188). This is not entirely true. A number of guenons, particularly the hardy patas monkey, are known to inhabit savannah regions. Only the monkey is shown in Vandier d'Abbadie's illustration, but Blackman's original drawing places the monkey on the lowest register of three bands of animals (1915: pl. 8). The top register depicts

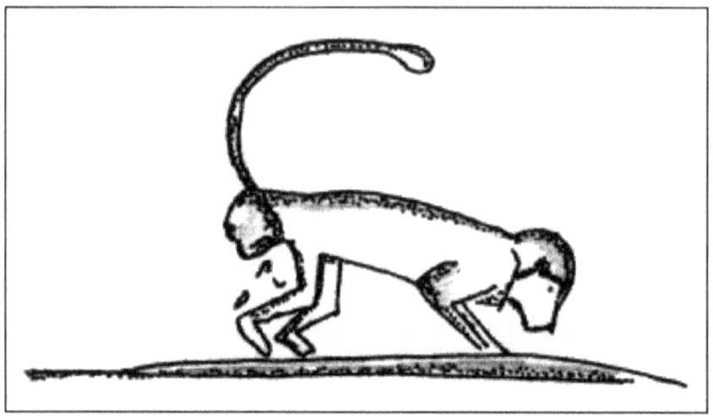

Fig. 28: A monkey gives birth
(After Vandier d'Abbadie 1965: 188, fig. 11)

cattle, including a mother and calf, and the second register depicts copulating gazelles and lions. The third contains a variety of animals that seem to be walking in a procession. The monkey stops to give birth. It seems probable that the scene is meant to depict a variety of environments, so the monkey might well be in a savannah setting rather than true desert.

THE MIDDLE KINGDOM: BABI AND THOTH

Although Hedjwer continued to be mentioned in spells from the Coffin Texts (Barguet 1986: spells 105,169, 1017), he was no longer a prominent individual deity. The baboon's religious significance became more closely associated with other gods, especially Thoth, with whom Hedjwer was already assimilated, and Babi. In many spells, the deceased adopts the powerful, aggressive, and sexual attributes of the baboon as personified by this god in order to protect himself from dangers in the afterlife. Babi was still the 'Bull of the Baboons' (Barguet 1986: spell 668), their great and terrifying leader with an enormous red phallus, which was equated with the mast of the ferry-boat in the realm of the dead (Barguet 1986: spells 397, 398). Its reeds were his slaver (Barguet 1986: spell 397). Spell 674, which gave the deceased the ability to live in the west and walk upright, ends with the affirmation, "He will not suffer, he will not be upside down, for N is Babi" (Coffin Texts, Faulkner, Trans. 1977). Spell 576, also gives the deceased the ability to continue sexual relations in the afterlife through the power of Babi (Faulkner, Trans. 1977):

> Copulating by a man in the realm of the dead. My eyes are the lion, my phallus is Babi, I am the Outcast, seed is in my mouth, my head is in the sky, my head is on earth…

Spells such as this illustrate an important aspect of the Egyptian conception of human and divine corporeality. Even divine bodies could be hurt and suffer death, and their ailments often mirrored those that most commonly afflicted humans. Both Horus and Re lose eyes, which are restored by Thoth, and Atum's eye was afflicted with a worm. However, with a few exceptions, divine bodies did not age and were essentially indestructible. The eyes of Horus and Re, for example, took on their own life. As Meskell notes, though, human beings could achieve such transformations only through language (Meskell and Joyce 2003: 15). By asserting that the phallus of the deceased is Babi, the member becomes divinized and separated from the body as a whole. It is endowed with the sexual power of the baboon.

Additionally, the god's bodily fluids, such as his saliva, can be considered self-sustaining substances, and any substance that issued from a divine body was considered productive (Meskell and Joyce 2003: 95). One might consider that the bark in the realm of the dead was also visualized as a body with separate, divinized parts, and both the mast and reeds were imbued with the power of Babi. In the case of the mast, one might argue that Babi's phallus was chosen partially because of the resemblance between the two objects. It seems significant that an aggressive god was chosen and that his secretion was saliva. The deceased's existence was in a precarious, liminal state as he was being ferried through the underworld. Perhaps the association of the parts of the bark with Babi's phallus and saliva lent a certain amount of security to the entire voyage.

The Osiris myth also grew in importance during the Middle Kingdom, and Horus and Thoth, because of their involvement in it, began to be represented on the east and west sides of coffins (Wilkinson, R. 1999: 71). An aspect of the conflict between Osiris and Seth that captured the Egyptians' imagination was the trial and the role of Thoth as advocate. Scenes of the judgment began to be depicted regularly on coffins, a practice that would continue into the New Kingdom (Boylan 1922: 25).

In Boylan's often cited study of the god, *Thoth: the Hermes of Egypt* (1922), literary references to Thoth from the Middle Kingdom are noticeably sparse perhaps because many spells from the Coffin Texts are practically identical to the Old Kingdom Pyramid Texts. A lack of material certainly cannot be to blame—Thoth is one of the most frequently mentioned gods in the Coffin Texts. Many of the spells, of course, make reference to Thoth's roles in the Osiris myth, and a brief study of these begins to create a more complete picture of the god's abilities.

Spell 228 mentions Thoth's role as scribe to Osiris. As in the spells concerning Babi, the deceased adopts the abilities and functions of the god. In spell 231, Thoth provides the gods and the deceased with speech through magic. This, along with his scribal function, became an attribute of the hamadryas baboon in the New Kingdom (Coffin Texts Faulkner, Trans. 1973).

A number of inscriptions from stelae erected at Abydos near the temple of Osiris also mention Thoth. In his autobiographical stele, Sehetep-Ib-Re, who served under Sesostris III and Amenemhet III, describes himself as 'straight and true like Thoth' because of his fairness in performing his duties (Lichtheim 1975: 126).

As in many other inscriptions, Thoth is linked with the highly praised quality of *maat*, which was personified by the goddess of the same name. In her physical form, she was depicted as a woman carrying an ankh and sceptre and wearing a stylized ostrich feather on her head. Although the concept of *maat* is usually translated as truth, this word cannot encompass all the nuances conveyed by *maat*, which represented the Egyptian view of ethical behaviour on both the human and divine levels. In early hieroglyphs, *maat* was composed of intersecting lines representing the king's throne, and it seems that geometric straightness or rectitude came to symbolize moral order. The goddess' association with the ostrich feather may have a similar origin because the feathers line up symmetrically along the quill (Armour 1986: 161-2). Thoth's wisdom in advocating gods and mortals who followed the path of *maat* led to an even greater relationship between the two gods by the New Kingdom, when the two were regularly depicted together both in solar and underworld contexts.

Finally, returning to the Coffin Texts, Thoth's role as lunar deity and his relationship to the solar god, Re, are mentioned in Spell 347 (Faulkner 1973):

> ...he will make me as Re and Thoth on that day and night when they were pleased(?) at it, Thoth performing the role of Re according to the command which Anubis made.

Thoth represents the night sky, while Re represents the day. He takes over Re's functions when the solar disk is on its nightly journey through the underworld. This theme, too, grew in importance in the New Kingdom, and it would be the baboon, rather than the ibis, which figured in lunar iconography.

THE NEW KINGDOM: NON-RELIGIOUS ROLES

From the Eighteenth Dynasty onwards, monkeys made their comeback in all forms of non-religious art. This was largely the result of the active colonization that took place after the departure of the Hyksos rulers. Ahmose I, the first ruler of the New Kingdom, made three successive expeditions (Vandier d'Abbadie 1966: 143). The expansion of Egypt's power to the southern lands led to the creation of numerous scenes of tribute bearers bringing baboons, vervets and other exotic animals to the Egyptian Pharaoh. During this period, when monkeys were portrayed with their handlers, these were usually young men from Nubia or Punt. Scenes with ordinary household servants became extremely rare.

A number of tribute scenes come from Theban tombs—those from the tomb of Rekhmire, who held the office of vizier in Thebes under Tuthmosis III and Amenhotep II, are particularly interesting. His title made him the highest civil official in the land, but he seems to have fallen from favour during the reign of Amenhotep II. His name was effaced throughout the tomb, and the burial chamber itself has not been found. The wall decoration, however, consists of over 300 square metres of both sacred and secular scenes and texts that provide valuable information about the duties of the vizier (Hodel-Hoenes 2000: 141-2).

The tomb seems to have been painted under the reign of Thutmosis III, during the period of Egypt's greatest expansion. The extent of the Empire is revealed in the number of different ethnicities depicted. Punt was probably the most remote land in the Empire, and, as mentioned above, it is believed to have been located in Sudan or Eretria. This, at least, is the origin of the products carried by the tribute bearers in Rekhmire's tomb: myrrh, a myrrh-tree, elephant tusks, a cheetah, leopard skins, ostrich eggs and feathers, hamadryas baboons and vervets. Directly following the southern tribute bearers are instructions for the vizier (Hodel-Hoenes 2000: 142):

> ...The state of the northern and southern fortresses is reported to him.
> Everything which leaves and enters the royal house is reported to him.
> Whatever enters into and exits from the region of the residence can enter or exit when his appointees allow it...

One of the most delightful scenes associated with the southern tribute bearers is depicted in the Nubian procession. A small vervet climbs the neck of a giraffe, who seems completely unperturbed. The giraffe's two conductors, however, seem to be looking on with a touch of astonishment (**Fig. 29**; Vandier d'Abbadie 1966: 153, fig. 12). Just ahead of the giraffe and its keepers are two other men leading monkeys. The first is carrying a large elephant tusk over his left shoulder and holding an adult male hamadryas by a leash. The proud male marches forward on all fours with a dignified air. The second man carries a basket of ostrich eggs covered with a large ostrich feather. He holds a leash attached to a lively little vervet that stands on its hind legs. One might guess that it is looking around for mischief, but its face is in lacuna.[6]

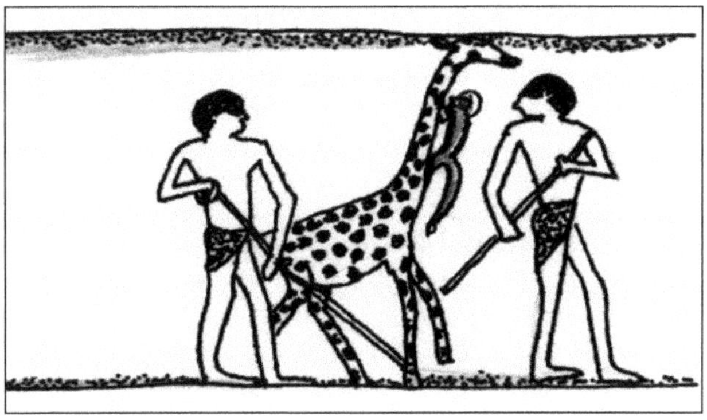

Fig. 29: A vervet climbs the neck of a giraffe in the Nubian procession from the tomb of Rekhmire

A similar scene from the tomb of Amunezeh depicts a vervet that has just jumped onto the back of a startled giraffe and is in the process of climbing its neck. Vandier d'Abbadie makes the apt observation that the monkey is treating its friend like the trunk of a palm tree (1966: 154, fig. 13). By the New Kingdom, monkeys had developed strong connections to various fruits and fruit trees. Both vervets and baboons were portrayed eating fruit or raiding baskets of fruit, and baboons, in particular became associated

[6] As a monkey lover, I cannot refrain from mentioning that these animals' leads are wisely attached to belts that they wear around their waists, rather than collars on their necks. The latter practice often results in accidental strangulation when the monkey tries to jump from any height without realising that the lead has become entangled around a fixed object.

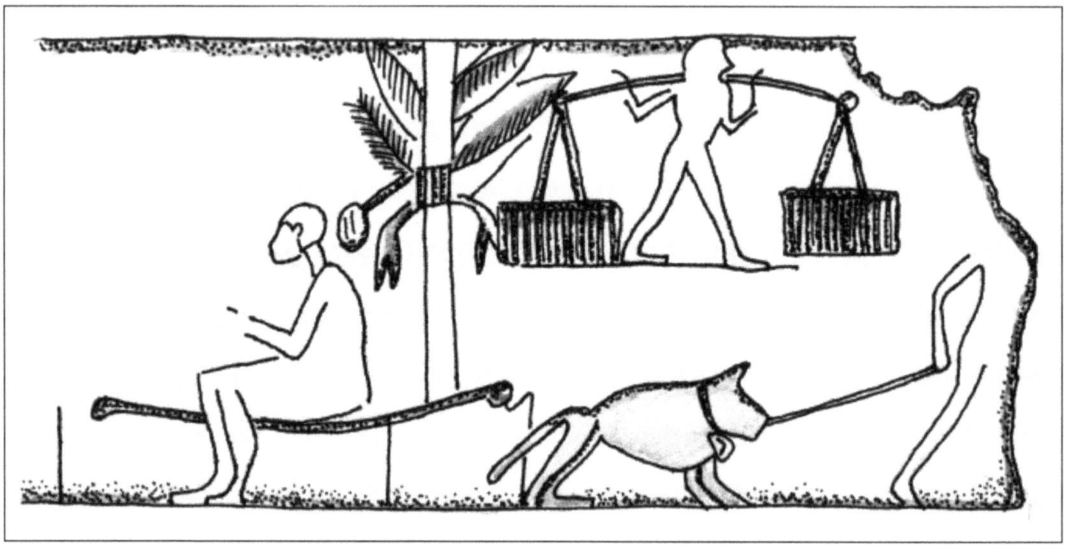

Fig. 30: A baboon refuses to be dragged away while dates are being picked
(After Vandier d'Abbadie 1966: 151, fig. 10)

with the dom palm tree. Such associations are depicted in a variety of media, including tomb chapel and temple walls, toilette implements, faience bowls, figurines and a large number of painted ostraca, most of which come from the workman's village of Deir el-Medina. A fragment of a wall-painting in the Berlin Museum humorously depicts the animal's attraction to dates. A young male keeper, partially in lacuna, is trying to force a hamadryas baboon away from a date palm tree. A man in the background is shown carrying away baskets of freshly picked dates, while another seems to be taking a break under the palm tree. The baboon stubbornly digs its hands and feet into the ground, apparently unwilling to leave while the harvest is still going on (Vandier d'Abbadie 1966: 151, fig. 10; **Fig. 30**).

Whether or not baboons were employed as fruit pickers, their connection with dom-palm nuts and trees was unmistakable by the New Kingdom.

The most common medium on which to depict this connection was figured ostraca, many of which were produced during the Ramesside period at Deir el-Medina. Vandier d'Abbadie catalogued over a dozen ostraca on which baboons are shown climbing the tall trees laden with large clusters of fruit. A number of these depict two baboons each climbing toward the clusters of fruit that hang from either side of the tree top. The baboons wear belts attached to leashes, and both animals are shown in profile, looking toward the right. On one ostracon of this scene type, a human figure is shown on the bottom right, holding one of the leashes (Vandier d'Abbadie 1937: nos. 2001, 2002, 2003). It is probable that a human keeper is implied in other such scenes. One limestone flake depicts a pair of male baboons flanking a dom-palm tree. The animals are almost as tall as the tree, and each rests one foot on a raised platform and the other on the ground.[7] The baboon on the right holds the trunk of the tree, while his companion on the left reaches up with both hands to grab a cluster of dom-palm nuts (Vandier d'Abbadie 1937: no. 2008). As in the Middle Kingdom tomb-chapel scene from Beni-Hasan, the actors in the scene are disproportionately large. It seems likely that each baboon is demonstrating a different gesture related to the nut gathering process. One may be shaking the tree, while the other pulls and twists at the clusters. Other scenes depict a single baboon climbing a dom-palm under the supervision of a Nubian boy (1937: nos. 2004, 2010).[8]

Ostraca are not the only medium in which baboons, and occasionally vervets, are connected to palm trees. An ivory spoon from the late Eighteenth Dynasty with a palm-shaped handle demonstrates the baboon's association with the tree, this time a date palm, as well as its association with unguents (**Fig. 31**).

Two figures are carved at the base of the tree. One seems about to climb the trunk, while the other leans back against it. Clusters of dates hang symmetrically on either side of the tree, and each is topped with a seated baboon. The animals face away from the trunk, and each holds a date to its mouth. Houlihan interprets the human figures as harvesters (Houlihan 1997: 43), but this interpretation does not seem entirely convincing. The leaning figure is holding a baton and seems to be unaware of the climbing nude figure. Manniche maintains that the man with the baton is a guard and the nude figure is a youth, who intends to pilfer some dates while the guard looks away (1999: 88). This seems a little more likely, although the nude youth has the curvy figure of a mature woman.

The monkey's connection to palm trees could also be indirect. A wooden kohl pot in the Louvre depicts a monkey, perhaps a baboon, sitting at the base of a palmiform column and hugging it. His head is turned to

[7] The baboon on the left is partially in lacuna, but the position of its legs seems to mirror that of the baboon on the right.

[8] It is important to note here that, although these representations of baboons and dom-palms do not seem to have any particular religious significance, Thoth as a baboon did acquire an association with the tree and fruit—an aspect that will be discussed further in the following section.

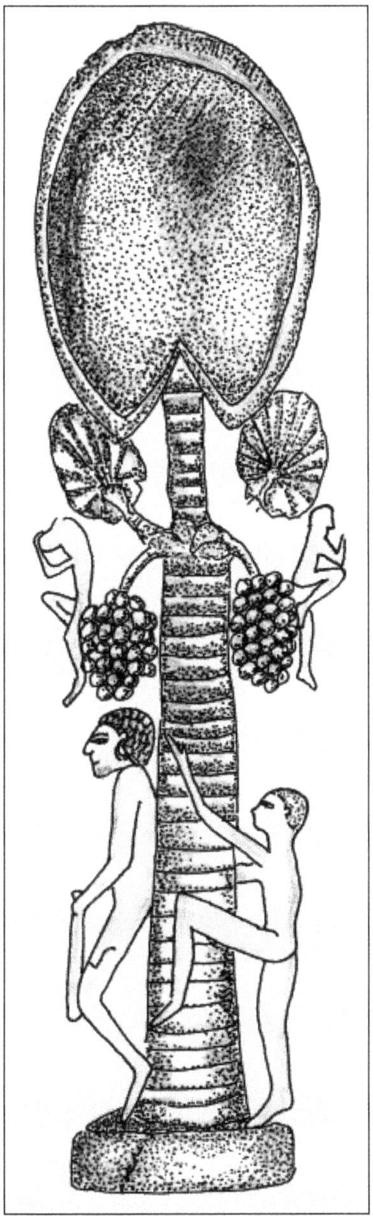

Fig. 31: An ivory spoon in the shape of a palm tree

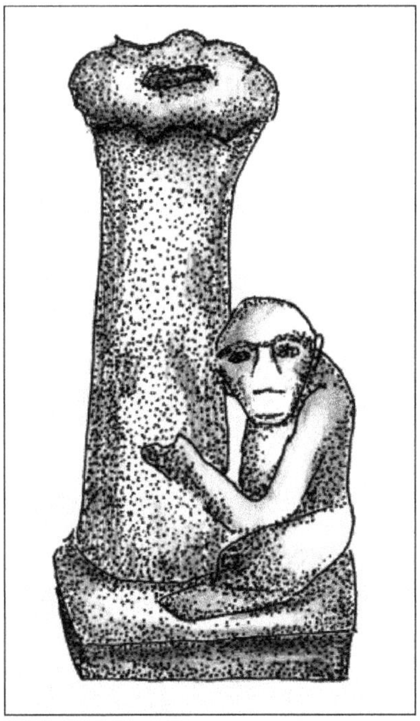

Fig. 32: A kohl pot in the form of a palmiform column held by a monkey (Louvre)

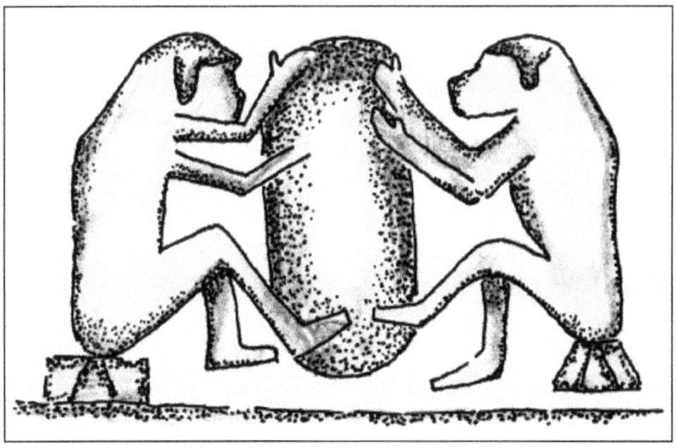

Fig. 33: Baboons eager for dom-palm nuts (After Vandier d'Abbadie 1966: 155, fig. 16)

the side so that the entire face is visible (**Fig. 32**; Vandier d'Abbadie 1972: 60, no. 183).

Elsewhere, baboons are comically shown tearing at sacks of dom-palm nuts in a desperate attempt to open them. An ostracon, again from Deir el-Medina, depicts a pair of baboons sitting on either side of a sack of palm nuts, tearing vigorously with both hands and one foot; the other foot is planted firmly on the ground for support. A third baboon, in lacuna, stands on all fours in the background (**Fig. 33**; Houlihan 1997: 41, fig 6; Vandier d'Abbadie 1966: 155, fig. 16). A number of other ostraca depict the baboons standing on either side of the sack, hitting it with large batons to break through the netting (Vandier d'Abbadie1966: 155).[9]

[9] From a personal communication with a primatologist at the University of Missouri-St. Louis, Pamela Ashmore, I have learned that hamadryas baboons in Awash still eat dom palm fruit, which leaves signs of wear on their teeth (30 Sep. 2004).

Vervets do not seem to have had an equally strong connection to palm trees, but they were occasionally shown climbing them. Their great fondness for fruit, however, was commonly portrayed. As a general rule, though, these smaller creatures are shown eating softer, smaller fruits than the dom-palm. One unusual wooden stele in the Egyptian Museum, Cairo, depicts a woman standing before a palm tree, perhaps a date palm, and holding a monkey's leash in one hand and basket of fruit in the other. The woman is eye-level with the clusters of fruit, and her monkey, who plucks one of the dates, is so enormous that its tail drags on the ground. Two pieces of fruit are lying on the ground under the tree. Vandier d'Abbadie argues that the woman is collecting the fallen fruit in her container and that the monkey is working for her to fill a small basket (**Fig. 34**; 1966: 197-8, fig. 57).

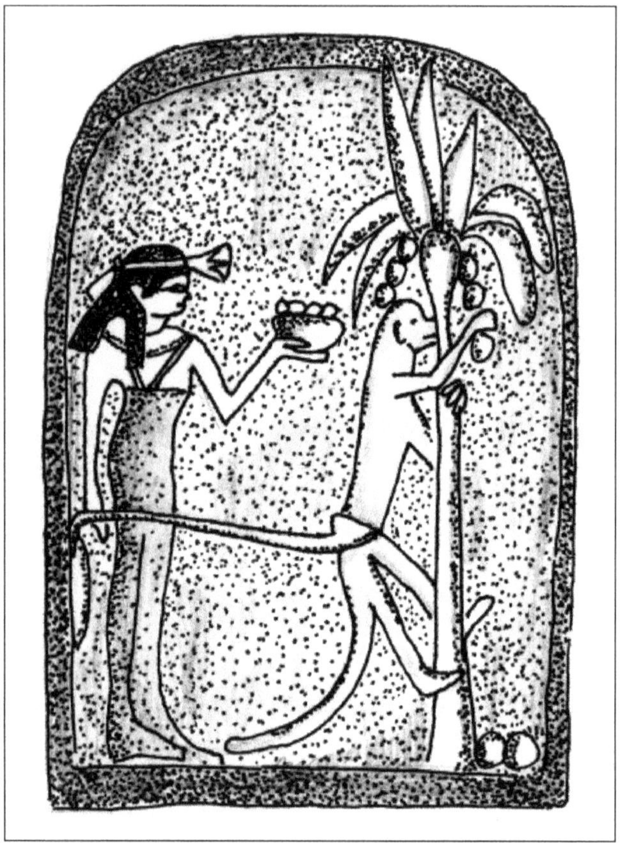

Fig. 34: A female harvester
(After Vandier d'Abbadie 1966: 197-8, fig. 57)

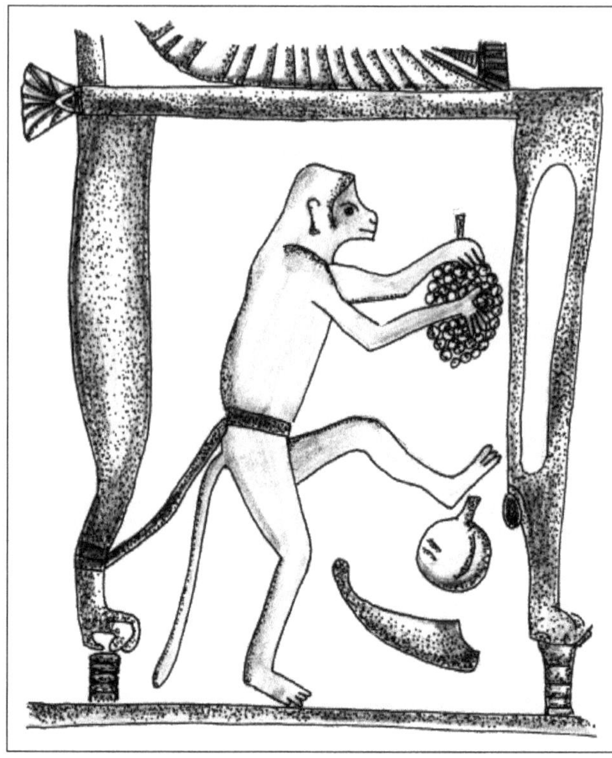

Fig. 35: A monkey with grapes dances for joy
(After Vandier d'Abbadie 1966: 165, fig. 24)

Houlihan, however, suggests the scene is comical in nature. He argues that the woman may be trying to persuade her pet to come down from the tree by offering it fruit from her basket (1997: 38). While I agree that this is not a typical harvesting scene because the monkey's handler is a woman and that it may, in fact, be a comical scene, it seems unlikely that the woman is trying to entice her monkey with the very fruit it is in the process of picking, especially since the animal is attached to a leash and could be pulled away from the dates. The monkey is probably a pet that has been allowed, perhaps even encouraged, to climb the tree by the woman, with the hope that her playful vervet will shake loose just enough ripe dates to fill her basket.

Numerous scenes of monkeys under their master's chair also depict these favoured pets eating fruit. On a stele in Cairo, a happy monkey is shown under its master's chair holding a big bunch of grapes with both hands and dancing with delight. The little vervet does not seem the least bit hindered by the leash that binds him to the chair (**Fig. 35**; Vandier d'Abbadie 1966: 165, fig. 24). Many scenes in which monkeys appear eating fruit actually depict the funeral feast. The monkeys are given a share in the form of fruit baskets or fruit that has fallen from a table of offering (1966: 168).

Comical scenes of monkeys sitting under the master's chair are also popular at this time. Strangely, though, the theme is not commonly associated with kings and queens. In fact, only one example of a monkey under a throne is known from the New Kingdom (**Fig. 36**; Vandier d'Abbadie 1966: 180). This is the amusing scene from the Eighteenth Dynasty tomb of Anen at Thebes. Beneath the throne of Queen Tiye, wife of Amenhotep III, a carefully rendered vervet jumps over two other pets, a cat and duck. The two seated animals seem to be the best of friends, as the cat protectively drapes a paw over his startled companion. Despite the lack of iconographic evidence, some kings did have simian pets. Tuthmosis III's baboon was found buried with him in the Valley of the Kings (Houlihan, 1996: 108), and at least one other king may also have had a pet baboon.

In 1906, Theodore Davis entered a pit tomb, KV 50, near that of Amenhotep II and discovered a scenario created by some ancient tomb robbers with a sense of humour: a large yellow dog standing face to face with a small seated hamadryas baboon (Davis 1908: 4-5; Ikram and Iskander: 27; Brier 1994: 211-12). Both mummified animals had been unwrapped and repositioned in ancient times, and Davis imagined the bright-eyed dog consoling his sad simian companion (1908: 4). It is possible, of course, that both pets belonged to the original tomb-owner, rather than the king.

Many painted ostraca or limestone flakes were also produced at this time with vervets or baboons and Nubian men or boys acting as their keepers, such as the flake in the Louvre (E 14340, Vandier d'Abbadie 1937: fig. 2038), which depicts a black man wearing the Nubian feather on his head and waving a baton at a baboon, who seems to be looking back at him with a bit of reluctance. The Nubians may have been baboon trainers, but it is impossible to know

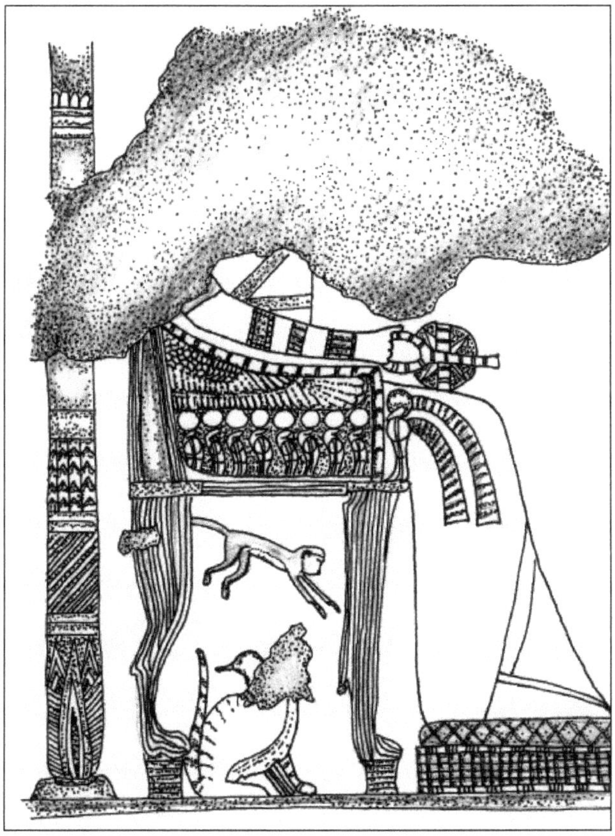

Fig. 36: A monkey jumps over a cat and duck in a scene from the tomb of Tiye

what they actually trained the baboons to do. Houlihan suggests the baboons may be young or recently imported animals that are being trained as house pets (1997: 42), but this seems a little improbable. The baboons appear to be adults, and a wild adult baboon, male or female, would not be an ideal house pet. Although some baboons were kept as pets, these must have been imported at a very young age and raised with the owner. Because the males tend to test their strength as they come of age and develop canines, one might well employ a baton to ward off an attack at this stage. However, because the baboons are walking and look back at the keepers in indignation, it seems probable that their keepers are trying to prod them along to a certain destination, perhaps to or away from palm trees.

Monkeys were also employed in a satirical fashion in variety of media during the New Kingdom. Ostraca were often used for practice sketches, and one, found in the mortuary temple of Ramesses II, which had been painted with the profile of a girl seems to have been mocked by a second artist who added a monkey to scratch her excessively large nose (**Fig. 37**; 2001: 93, fig. 101).

A unique Eighteenth Dynasty scene may parody the common representation of the goddess, Nut, in a Sycamore tree, pouring pure water for soul birds who receive the liquid in their hands. Here a baboon, whom Vandier d'Abbadie maintains is completely devoid of religious significance, seems to imitate the gestures of the soul-birds in an irreverent manner (**Fig. 38**; 1966: 160, fig. 20).

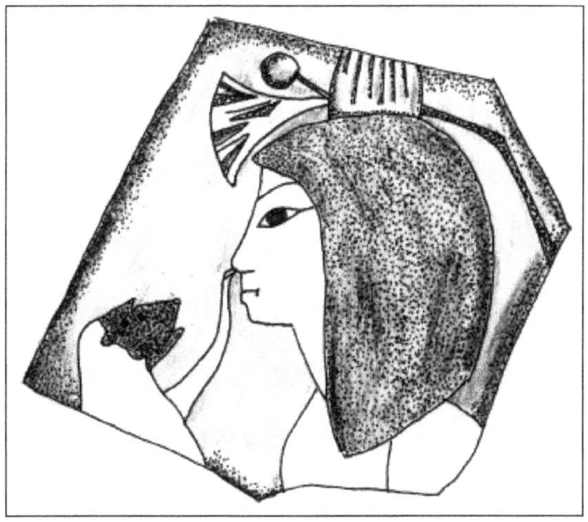

Fig. 37: A second artist mocks the practice sketch of a girl with a large nose (After Houlihan 2001: 93, fig. 101)

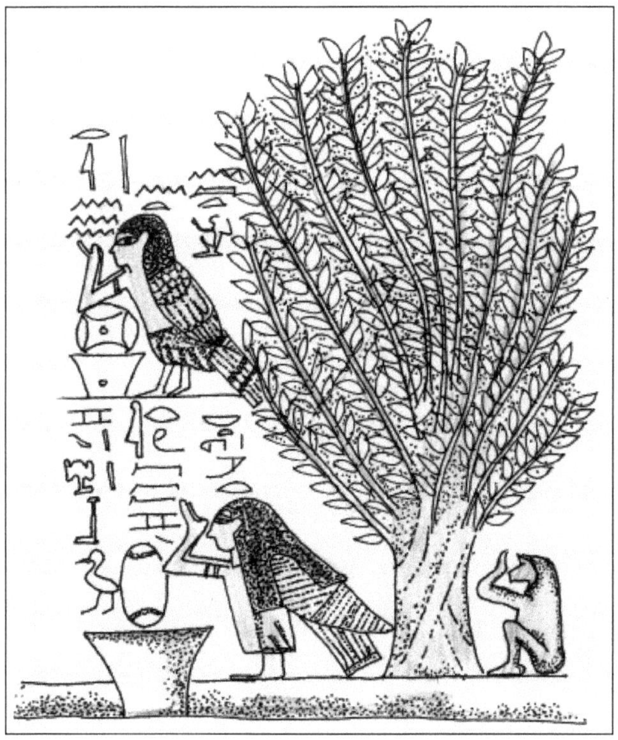

Fig. 38: An irreverent gesture? (After Vandier d'Abbadie 1966: 160, fig. 20)

Thoth, however, did have an association with Nut, so it seems possible that the god is present in his baboon form, lending support to the deceased.

Political satire may also be found in some of the limestone monkey statuettes and reliefs found by Petrie at Tell el-Amarna. Although most of these objects functioned as toys, a few of them also seem to be making a monkey of the heretic king, Akhenaten. One painted limestone toy depicts a monkey driving a chariot pulled by another monkey. Both the front and back wheels have holes drilled into them to allow the insertion of a wooden spoke for moving wheels.

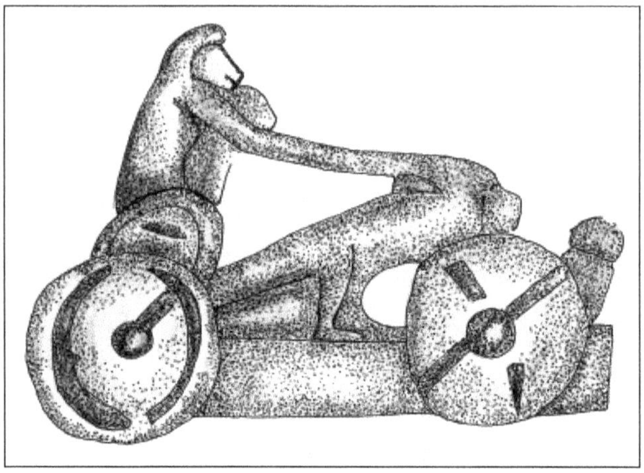

Fig. 39: An artist's impression of Akhenaten?
(After Houlihan 2001: 64, fig. 58)

It has been suggested that the simian charioteer represents Akhenaten, who regularly rode through the city with his family (**Fig. 39**; Houlihan 2001: 63-4, fig. 58; Frankfort and Pendlebury 1933: 99, Pl. 31, fig.4).

Another fragmentary statuette depicts a baboon driving a horse-drawn chariot, which has a hole for the addition of wheels (**Fig. 40**; Samson 1978: 37, fig. 16). It seems likely the baboon's large head and long muzzle reminded the artist of the peculiar facial features with which Akhenaten had himself portrayed. Finally, an unfinished statuette of two monkeys grooming each other may be a parody of the scenes depicting the king kissing his daughters (Samson 1978: 40; Frankfort and Pendlebury 1933: 99, Pl. 31, fig. 8).

Three satirical papyri, probably from Deir el-Medina, depict comical scenes of a topsy-turvy world in which cats serve mice and lions play board games with animals that would normally be their prey. Monkeys and other animals play musical instruments for a mouse queen. These papyri are housed in London, Turin and Cairo, and Brunner-Traut describes them as "the oldest 'comic strips' of world history" (1979: 11). In addition to their comic value, the scenes may represent an earthly paradise or the chaotic period of the New Year Festival, which has far greater documentation in Mesopotamia (Brunner-Traut 1979: 14).

The greatest source of these humorous scenes can be found on the figured ostraca, many of which were also produced by the skilled workmen of Deir el-Medina. This village was unique in that an astounding forty percent of its population was literate during the height of its prosperity in the Twentieth Dynasty (Houlihan 2001: 61). Numerous ostraca show monkeys dancing or playing musical instruments. One limestone flake depicts a monkey playing a boat-shaped harp for a mouse dressed as a nobleman, while another bears the figure of a baboon playing the double flute for a second baboon dancing with batons (Houlihan 2001: 89-90, figs. 94, 97).

The Egyptians were equally aware of the hardships wild monkeys suffered as a result of human behaviour. A scene

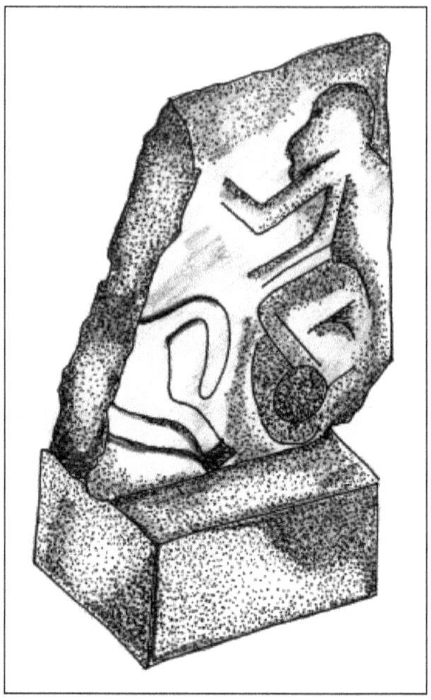

Fig. 40: Another toy depicting Akhenaten as a baboon.
(After Samson 1978: 37, fig. 16)

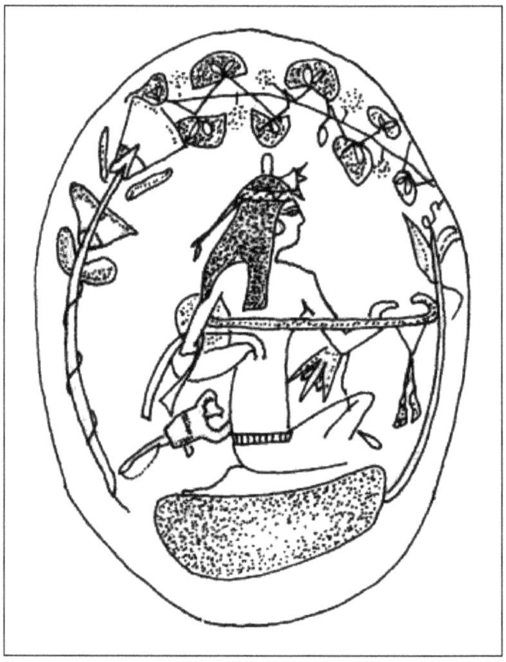

Fig. 41: A blue faience bowl of a girl and her monkey
(Boston Museum of Fine Arts)

from the temple of Medinet Habu depicts a massacre of Nubians by the army of Ramses III during a military campaign in the country. In a corner of the wall, a woman is shown crouching below a palm tree with a basket, while a frightened monkey climbs the tree to escape the dangers of battle (Vandier d'Abbadie 1966:197).

Monkeys also continued and expanded their association with women and containers in the New Kingdom. A blue faience bowl from the Nineteenth or Twentieth Dynasty depicts a woman, nude except for jewellery and a girdle, sitting on a cushion and playing a musical instrument. Her thigh is tattooed with an image of the fertility god Bes, and a monkey stands behind her, tugging on her girdle. The erotic nature of the scene is further implied by the presence of stylized cowrie shells, representing female genitalia (**Fig. 41**; Friedman 1998: 212). This may be viewed as an early association of the monkey with Bes. As previously mentioned, by the Third Intermediate to Late periods, the monkey would commonly be found on amulets seated at the dwarf god's feet, perched on his shoulder or cradled in his arms (Dasen 1993: 57; Romano 1989: 133; Bulté 1991: 80-82). Many cosmetic containers and hairpins also incorporate monkey figures at this time, and Manniche notes that some graffiti of sexual intercourse depict women with monkey-like faces (1997: 43-4, figs. 32, 34). Clearly, then, monkeys were strongly linked to female sexuality both indirectly and directly by the New Kingdom.

THE NEW KINGDOM: BABOONS AS DEITIES AND SUN GREETERS

By the New Kingdom, baboons had developed strong connections to the moon through their association with Thoth and yet another lunar god, Khonsu, whose origins are uncertain, but he is known from the Old Kingdom tomb chapel of Pepy II. His usual form is that of a mummified young man or child, wearing the sidelock of youth and a crescent or full moon on his head. His name derives from a word meaning 'wanderer' or 'to cross over,' and his name appears to mean 'he who traverses,' an appropriate lunar title. By the end of the Middle Kingdom, he had become part of the Theban triad, Amon-Mut-Khonsu. Previously, he had been associated with the mummified falcon, Sokkaris (Lachaud 1995: 91).

In the New Kingdom, he became closely associated with Thoth, who, by this time, was considered the second Re. Thoth was the nocturnal double of the solar god, who allowed the sunlight to remain even through the darkness. Khonsu's function corresponded more to the phases of the moon and the rhythm of the tides (Lachaud 1995: 92). His form became conflated with that of Thoth, and blue faience figurines of baboons wearing the lunar disk may depict either god (**Fig. 42**). The ibis was also represented in this colour, but it is not known precisely why the colour blue was used. The colour is often associated both with the heavens and the primeval flood, and solar objects are often painted blue (Wilkinson, R. 1999: 107-8). Silver, the metal associated with the moon, was often added to figurines to enhance the lunar connection (1999: 84). In some cases, the baboon may also hold the *wedjat*, the mooneye (literally, the restored or sound one) of the solar falcon, Horus. These can clearly be identified with Thoth, who was credited with its rescue in the Pyramid Texts.

By the Eighteenth Dynasty, the baboon's solar connections were equally strong. A New Kingdom hymn describes their sun-greeting function (Houlihan, Trans. 1996: 96):

> The baboons that announce Re when this great god is to be born again about the sixth hour in the netherworld. They appear for him after they have come into existence. They are at both sides of this god until he rises in the eastern horizon of the sky. They dance for him, they jump gaily for him, they sing for him, they sing praises for him, they shout for him. When this great god appears before the eyes of [all humankind] than these hear the speech of jubilation of the *Wetenet*-country [=baboons?]. They are those who announce Re on heaven and earth.

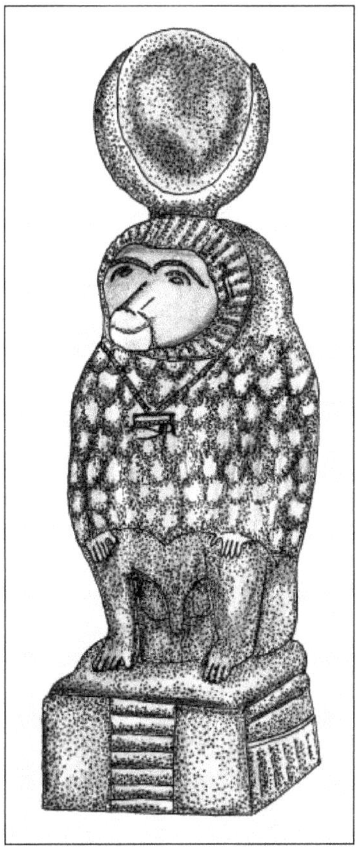

Fig. 42: A blue faience statuette of Thoth (Louvre)

This hymn is reminiscent of the Pyramid Text Utterances in which the sons of Re are deified baboons. It is generally believed that the Egyptian view of baboons as sun worshippers may have come about from observing the animals in the wild as they gather together at sunrise and chatter loudly. In reality, they are discussing the direction to take in search of breakfast, and captive baboons are simply chattering and stretching. The Egyptians, however, viewed the hamadryas as the most religious of creatures, endowed with a sacred language that could only be understood by the pharaoh (te Velde 1988: 134). Te Velde also remarks that the jumping and singing of the baboons in the sun hymn may denote ecstatic behaviour and that the portrayal of male baboons as ithyphallic in their adorant attitude with raised forearms may be viewed as iconographic reinforcement of their religious ecstasy (**Fig. 43**; 1988: 130-1).

It is difficult to know whether the belief that baboons possessed a secret language was the result of their association with Thoth or whether Thoth was considered the special patron and protector of scribes in his baboon

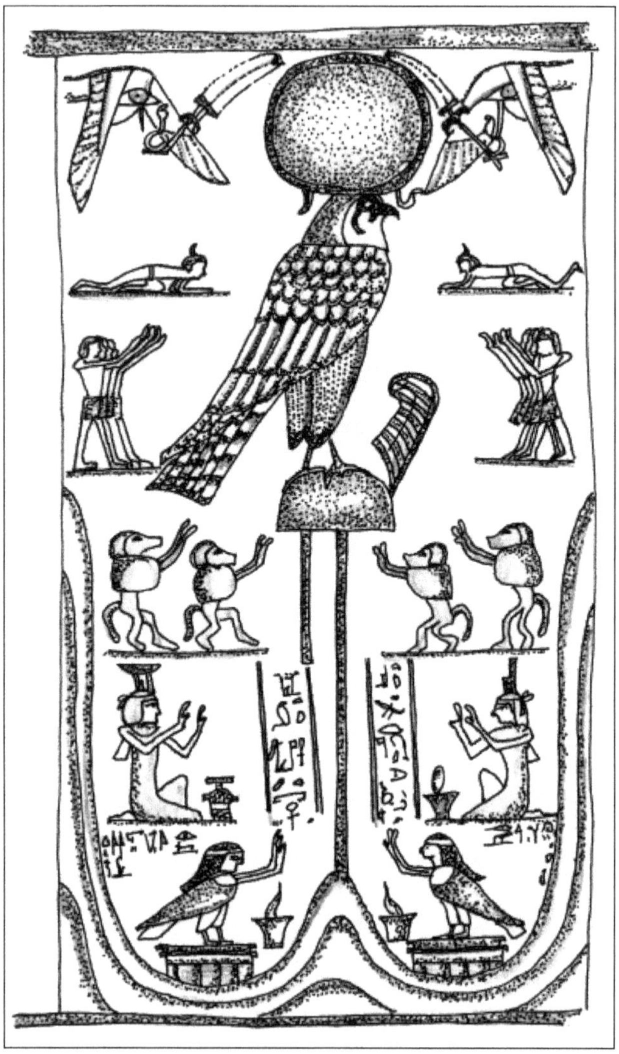

Fig. 43: A vignette depicting baboons greeting the sun in the form of a falcon wearing the solar disc (BM 10472/1)

form because of the Egyptian's perception of the hamadryas a particularly clever and vocal animal. In any case, numerous small monuments from the New Kingdom portray scribes sitting with Thoth in his baboon manifestation to demonstrate their closeness and importance to the god. One Nineteenth Dynasty statuette from Hermopolis, now in the Ashmolean Museum, depicts a standing scribe, wearing the leopard skin associated with priests. Thoth in his baboon manifestation sits on the scribe's shoulders and clings to his head. The base of the figure was broken off, but cartouches are visible on the front and back of the figure's left shoulder. Hieroglyphs on the back pillar indicate that the statuette was a funerary offering to the individual, who was a priest as well as a scribe (Museum of Fine Arts, Boston 1982: 282, fig. 387). The statuette belongs to an unusual group in which the human figure is not portrayed making any scribal gesture. Santolini maintains that Thoth as a baboon is never portrayed holding scribal equipment or making any scribal gestures. He seems merely to supervise and protect the individual (1984: 215). In sculptural representations, this is perfectly true, but in one vignette of Spell 125 from the Book of the Dead from the papyrus of the Royal Scribe and Chief Military Officer, Nakht, the deceased kneels before Thoth, who is present in his baboon manifestation, wearing the lunar disc and holding a scribal palette (BM 10471).

Two of the most famous and beautiful examples of a more common group are the small monuments, one of alabaster, the other schist, of the prince and royal scribe, Nebmertuf, who appears beside Amenhotep III in a series of bas-reliefs. They are believed to have come to the Louvre's collection from the temple of Hermopolis, Thoth's main cult centre, where the same pharaoh had erected four colossal hamadryas statues to him (Andreu 1997: 21). Each portrays Nebmertuf sitting pensively, leaning over the papyrus before him, seemingly unaware of the handsome male hamadryas sitting above him on an altar to his left. The schist seems a better medium to capture the features of the young prince and stately baboon, whose muzzle, unfortunately, has been broken off.

The monument bears three inscriptions. Those around the base and on the papyrus name Nebmertuf and his functions, while the one circling the altar reads (Personal trans. from French, Bénédite 1911: 13):

> Living Thoth, Lord of Ashmunein (Hermopolis), great god in Heserit, king of Heaven, guide...who restores the lunar eye to his master.—Living Thoth, master of divine speech, author of the truth each day, who presents it on his two hands to the Sun, who rejoices in it each day.

The inscription links Thoth to Maat, truth, and the restoration of the lunar eye to Re. As previously mentioned, Thoth is Re's double. A prayer from the Book of the Dead mentions honouring the Ba at the sixth hour and looking upon the fifteenth day of transformation. The sixth hour symbolizes the crescent moon and the fifteenth day the full moon. The full moon represents a mirror in which the sun is reflected in the night, and its appearance guarantees the return of the sun at dawn (Lachaud 1995: 159). Thoth is also credited with restoring one of Re's eyes, either solar or lunar, in a variety of myths, and the expression "heart of Re" became a common epithet of Thoth (Santolini 1984: 212).

Artistically, Thoth's baboon never looked more alive, realistic or benign than during the reign of Amenhotep III. Like the smaller statues just described, the colossal quartzite baboons at Hermopolis, which weigh around thirty-five tons, were also carved with considerable attention to detail (**Fig. 44**).

The most beautiful example I have come across is the brown quartzite statue of Thoth, inscribed with the name of Amenhotep III, in the British Museum. The baboon's serene face is perfectly rendered, and the striations of the fur around the animal's face realistically cover the ears while conforming to the contours of the head. Some stylization is visible in the symmetrical strands of fur that make up the cape, but the overall effect is stunning. Attention was also devoted to the back of the statue, where the baboon's bare bottom is visible just below the shorter fur of his lower back. The artist even rendered the tuft of fur at the end of the hamadryas' tail by increasing the width of the

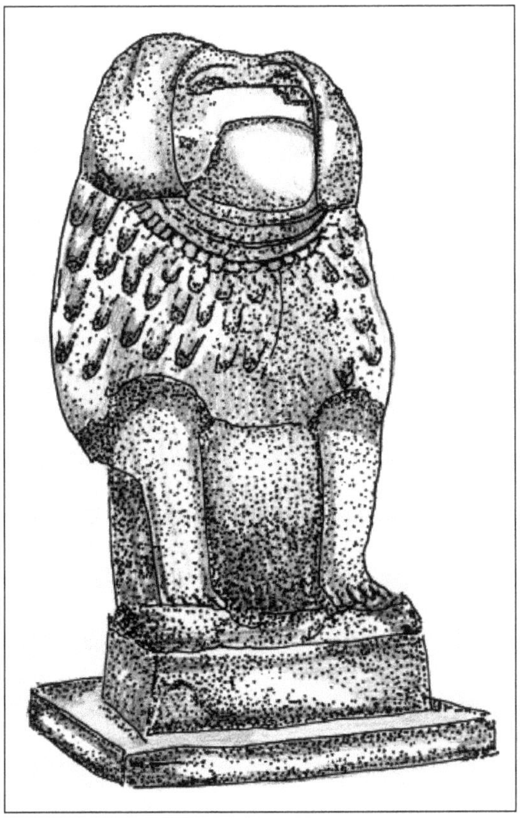

Fig. 44: A colossal baboon still in situ at Hermopolis

The baboon also gained importance in funerary iconography during the New Kingdom. One of the four sons of Horus represented on canopic jars, Hapy, who protected the lungs of the deceased, came to be depicted with a hamadryas head in the Nineteenth Dynasty, while even earlier vignettes from Spell 125 of the Book of the Dead depict Thoth as an advocate of the deceased recording the result as the heart is weighed in the balance. Some vignettes reinforce Thoth's presence by showing him both in his anthropomorphic ibis-headed form and as a baboon perched on the balance (**Fig. 45**). As previously mentioned this god knew the original sacred language of the gods and could function as an interpreter in the netherworld (te Velde, Trans. 1988: 136):

> I am Thoth and I speak to you the language of Re as herald. They spoke to you before my words were understood. I am Thoth, the lord of divine speech, who puts things in their right place. I give god's offerings to the gods and invocation offerings to the blessed dead.

Thoth was, therefore, an intermediary between the human and divine realms who aided Egyptians in their most important transitional phase, the rite of passage from this world to the next.

In his ground breaking study, *Les rites de passages*, Van Gennep devoted a chapter to the funeral ceremonies of various cultures, which often include a period of time

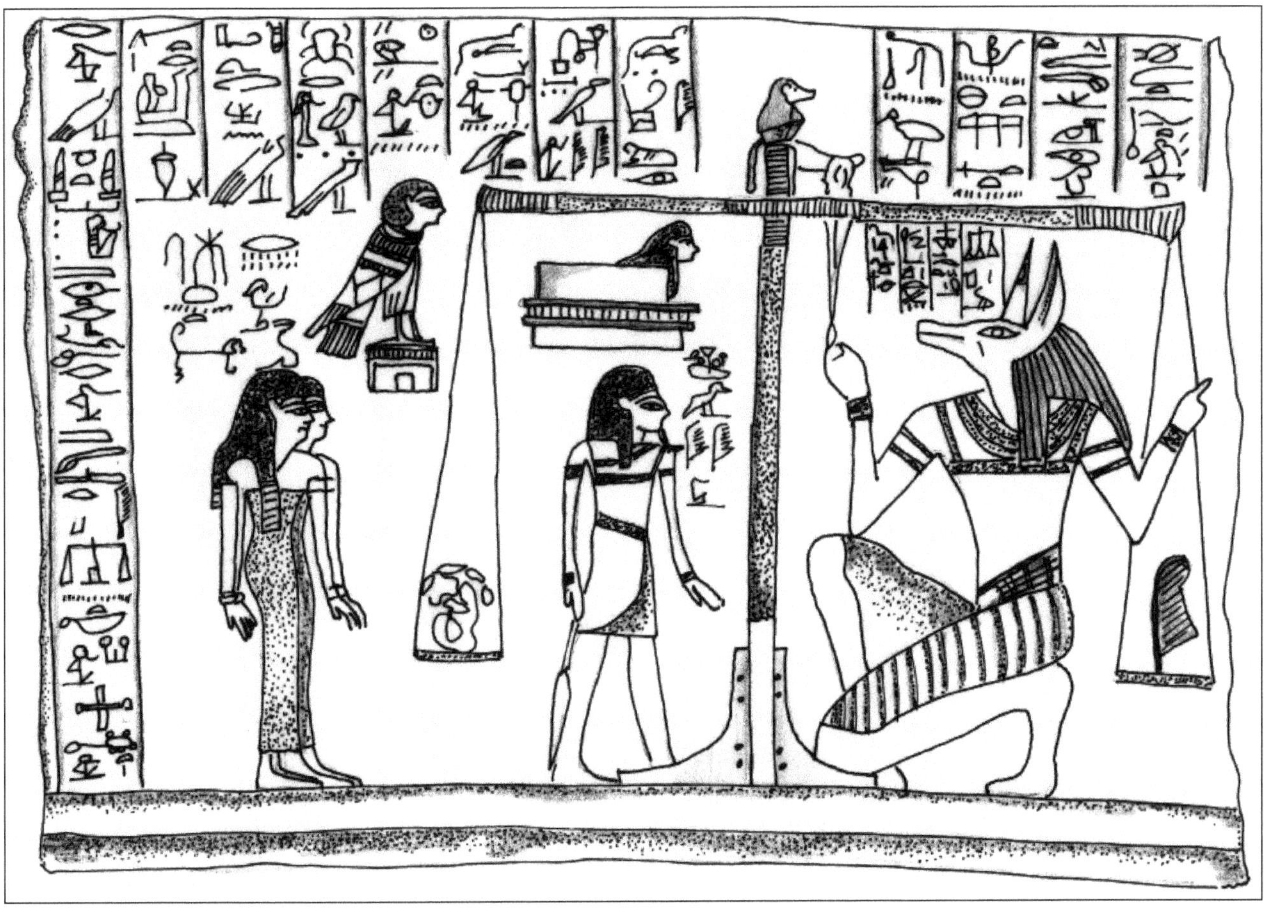

Fig. 45: A vignette of Spell 125 from the Book of the Dead from the Papyrus of Ani

when the deceased is believed to exist in a liminal state between life and death and must go through a series of rites of passage before being accepted in the realm of the dead. In Egypt, this was considered a time of great danger—if the deceased was not pure, his heart would be devoured by the monster, Ammit. This, in effect, would be a second and final death. The belief that the deceased would have a powerful and just god, such as Thoth, as his advocate during this process alleviated many of the fears associated with death. In many cultures, after this transitional period, a means of transport is required to carry the deceased to his final destination (1909: 153). The Greeks had Charon to act as ferryman, and Hermes also played the role of psychopompos. In the Old Kingdom, Thoth as an ibis assisted the deceased king in this journey by carrying him on his wings. It is easy to understand, then, how Thoth would later become conflated with Hermes in the Greco-Roman Period.

In his status as intermediary, Thoth also acted as an arbitrator between the gods. Two ostraca from Deir el-Medina, dating to the Ramesside period, relate the myth of the goddess, Hathor-Tefnut, who, when she became angry with her father, Re, fled to the Nubian desert and transformed herself into a cat. At Re's request, Thoth journeyed to the desert and, after transforming himself into a baboon, persuaded the goddess, through a series of fables, to return to Egypt (Janssen and Janssen 1989: 60, fig. 52; Vandier d'Abbadie 1966: 158, fig. 19).

A strange New Kingdom story of gods behaving badly also depicts Thoth as a necessary intermediary and advocate. When Horus is sexually assaulted by Seth and catches the semen in his hand, Isis immediately chops it off in disgust. She then visits Seth's garden and pours Horus' semen onto the lettuce. Thoth becomes an accomplice in the revenge of maligning Seth as a semen eater by pulling Horus' semen from Seth's head in the form of the solar disk. The Ennead, impressed by the display, declare Horus right and Seth wrong (Simpson, trans., 1973: 108-126).

The baboon's fondness for dom-palm nuts also led to a connection between Thoth and the dom-palm tree and fruit. Santolini explains that appearance of the tree itself is mathematical. Its trunk forks into sections, and the fruit, which grows in clusters, is ovoid and composed of edible inner and outer parts. The Egyptians eat the outer fruit of the dom-palm nut, but the albumen in the centre is accessible only to baboons, who are able to crack the casing with their powerful jaws. If the nut were cracked open by a human, most of the liquid would be lost (Santolini 1984: 214), so it is natural that the god of magic, wisdom and calculation would be interested in such a complex tree and the mysterious nectar contained in the fruit. In a hymn to Thoth, the supplicant asks for the water in the interior of the palm nut, and Santolini maintains that it is not a physiological thirst that could be quenched by this liquid. He views the dom-palm as the issue of Thoth's science and the liquid at the centre of the nut as fluid that sustains him through his baboon manifestation. Essentially, when the baboon eats the liquid, it is as if Thoth is assuring his own continuity. Additionally, both Thoth and the dom-palm nut were believed to possess a certain sexual potency. Part of the nut bears a resemblance to genitalia and is believed to have been valued by the Egyptians for this similarity, and Thoth, in his baboon manifestation, was generally portrayed as ithyphallic (Santolini 1984: 216-7).

Babi was still an active and fierce presence in the New Kingdom. The Jumilhac papyrus portrays him as a dog, but Derchain maintains that the description of his eyes is more reminiscent of a baboon (1952: 28). He is often conflated with Seth by this time (1952: 34-5). In the story of Horus and Seth, Babi manages to deliver such an insult to Re-Harakhti that the latter is compelled to lie on his back with a sore heart. Babi is then chased off by the Ennead. The address to the gods from the Book of the Dead Spell 125, even contains a plea from the deceased to be rescued from Babi, who feeds on the entrails of nobles. The god's aggression clearly became more pronounced by the New Kingdom, and he may be viewed as an irrational, violent counterpart to Thoth, who is always an enemy of chaotic forces. An unfinished wall painting from the Book of Gates in the tomb of Horemhab depicts Thoth in his baboon form brandishing a baton to drive away the pig of Seth (**Fig. 46**; Houlihan 1996: 27, fig. 21).

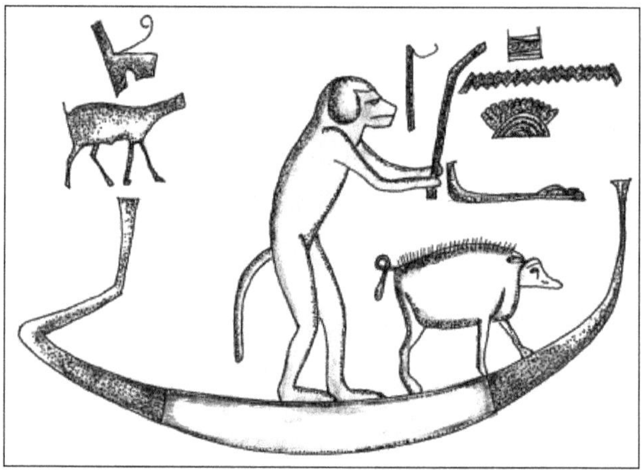

Fig. 46: Thoth as a baboon drives away Seth (After Houlihan 1996: 27, fig. 21)

As previously mentioned, the behaviour of the hamadryas baboon may be largely responsible for the conflicting portrayal of two gods with baboon manifestations. The hamadryas is known to cultivate female baboons at a very young age as future members of his harem, while other species show no interest in females before sexual maturity. Males can be extremely charming in their attempts to attract the girls, lip-smacking and making low, coaxing sounds, but surprisingly violent in keeping members of the harem from cavorting with other males.[10] Any Egyptian familiar with the animals must have noticed these sudden changes of temperament. They are strongly linked to the baboon's sexuality, and all portrayals of baboon deities, including Thoth, represent the baboon with his prominent phallus. Thoth's baboon may have been inspired by

[10] From my personal experience with baboons, I can say that this extreme jealousy extends even to familiar human females. I once received a quick but serious warning bite from my hamadryas, Herbie, when a male friend of the family touched my arm.

the image of the older, kinder male hamadryas, who demonstrates social success with clever stratagems rather than angry outbursts. Babi behaves more like a youthful baboon, who attempts to demonstrate his strength through aggression. The Egyptians' desire to control this negative aspect of the animal and use it as a form of protection is also evident in the depiction of other aggressive animals as protective deities, such as the hippopotamus goddess, Taweret, described above.

Other baboons in the religion of the New Kingdom also maintained this aggressive, protective function. The western wall of the sarcophagus chamber of Tutankhamun is decorated with an extract from the Book of the Amduat, which depicts twelve ithyphallic hamadryas baboons representing the first hour of the night. These are the beneficent genii, known as turquoise baboons (**Fig. 47**; Beler 2001: 67).

Vignettes and spells from the Book of the Dead also depict protective baboons. A vignette of spell 126 from the papyrus of Ani depicts four hamadryas baboons sitting around a Lake of Fire (**Fig. 48**). Flaming braziers separate the animals. The lake is not mentioned in this spell, but these are the baboons who sit in the bark of Re and allow the deceased to enter the portals of the West, where he will be given sustenance. In spell 125, the deceased must declare his innocence before the gods of the tribunal. One, known as Nosey of Hermopolis, may be a baboon associated with Thoth, and indeed some vignettes depict a number of these deities with baboon faces. Because spells from the Book of the Dead continued to be used well into Greco-Roman times, the baboon never lost these important protective functions.

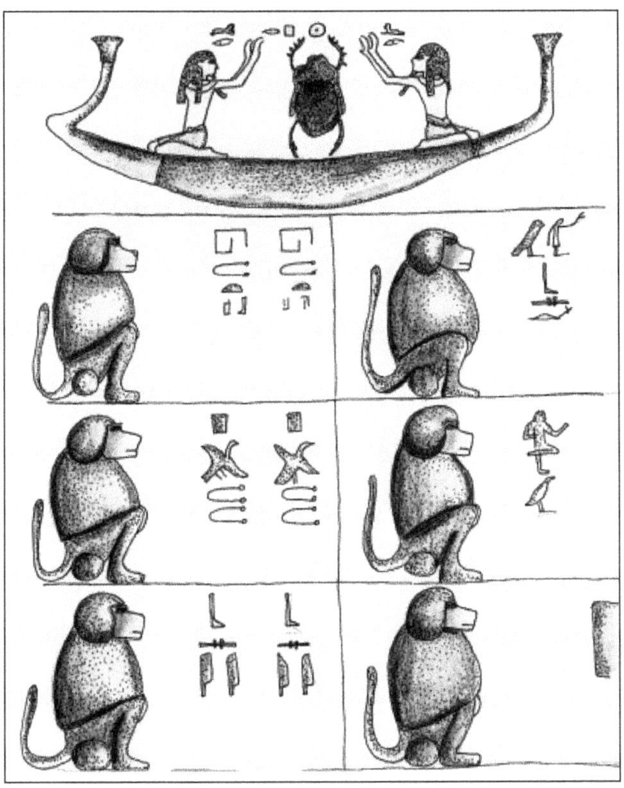

Fig. 47: Baboons on the wall of Tutankhamun's sarcophagus chamber (After Beler 2001: 67)

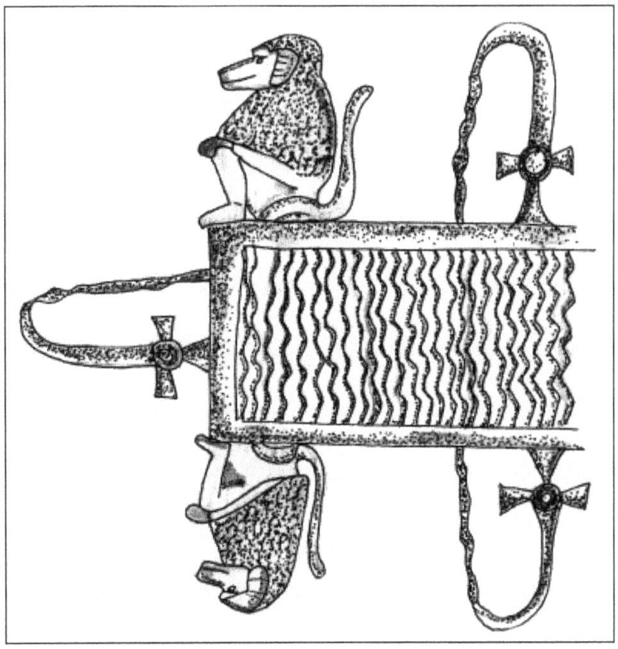

Fig. 48: A vignette from Spell 126 from the Papyrus of Ani

THE THIRD INTERMEDIATE PERIOD TO THE PTOLEMAIC PERIOD

After the New Kingdom, monkeys more fully developed their association with women's fertility amulets. Numerous small plaques depict monkeys and other animals sitting on or being held by Bes or a voluptuous nude woman with an exotic hairstyle, while some amulets represent monkeys without these figures. These blue-green faience objects have never been securely dated, and estimates have ranged from the Third Intermediate Period to the Ptolemaic period. Bulté, however, recognised a similarity in colour and substance to faience of the Libyan period, and most of the figurines of secure contexts were found in the eastern Delta, where Bubastis and Tanis flourished during the Libyan period (Yoyotte in Bulté 1991: 8). The monkey depicted is always one of the smaller species, most likely a vervet. In Bulté's study of the known blue-green faience plaque-amulets, she identified three main thematic groupings that included figurine-plaques of Bes, vervet figurines that may have had connections to Thoth, and the female figurines linked to Bastet. The groupings are related to a number of iconographical themes in scenes that depict the nurturing of new-borns, music and animals, toilette and plants, including reed, rosette and papyrus-shaped objects, palm trees and dom-palms (Bulté 1991:11).

She identifies six variations among the amulets depicting monkeys. The first is a seated vervet, probably female, holding a papyriform rosette. Smaller monkeys may be present on the amulet. One example in the Cairo Museum depicts a large vervet holding a circular object surrounded by a rosette, which is supported by a cluster of smaller monkeys seated on a palm trunk. Two other monkeys at her feet cling to a large palm leaf (Bulté 1991: 35, Pl. 17a, JE 72520). In the second variation, a seated monkey holds two smaller monkeys, seated back to back, with their hands held up to their mouths. A third monkey seated at her feet also makes this gesture (**Fig. 49**; 1991: 37, Pl. 16a, JE 31670). The next variation comprises a vervet holding her palms to her mouth as if eating fruit. Her infant, facing left, sits before her, making the same gesture. The fourth variation depicts a walking vervet, while the fifth is a vervet musician, who may also be eating fruit. A single example is known of the sixth variation, in which five monkeys climb the trunk of a tree, while antelopes rest at the base (Bulté 1991: 39, Pl. 17e).

The amulets that depict Bes or a female figure with vervets are most likely designed to protect new-borns. It seems the vervet was also associated with Atum at Heliopolis at this time (Bulté 1991: 99). In support of the monkey's solar connections, Bulté notes that Thoth adopted this form of monkey, *gf*, when he prepared the rites of appeasement for Hathor-Sekhmet. Additionally, the Second Book of Breathings invokes the power of Re, under his aspects of Baboon of Bahou and Vervet, *gf*, of Kush, to give guidance to the new-born. She remarks also that the monkeys are consistently associated with dom-palm nuts that contain Thoth's self-sustaining fluid (1991: 100). The rosette decoration around the circular object held by the monkeys seems to be a papyrus plant, which was used as a decorative motif on a number of toilette objects related to the New Year festival and birth. The function of the solar object is unknown, but it may have been a solar symbol or even a tambourine (1991: 101).

Male sexuality was also protected by monkey amulets. A faience monkey, dating from the Twenty-Sixth Dynasty to early Ptolemaic period, in the Leo Mildenberg collection represents a common amulet that was placed in the coffin with the deceased to ensure his sexual powers in the afterlife. The monkey, probably a vervet, walks upright, displaying his genitalia. His arms hang by his sides (**Fig. 50**; Walker 1996: 68, fig. 96). It is worth noting that vervets often stand on their hind legs during social interactions, which allows the males to display their bright blue testicles and red penis (Kavanagh 1983: 142). The Egyptians, who commonly kept these little simians as pets, must have been aware of their behaviour and probably modelled the amulets accordingly.

Another Late Period amulet that remained popular even in the Ptolemaic period was the hypocephalus. This circular wooden amulet, inscribed with spell 162 from the Book of the Dead, was placed beneath the head of the deceased with the intention of causing a flame beneath the head

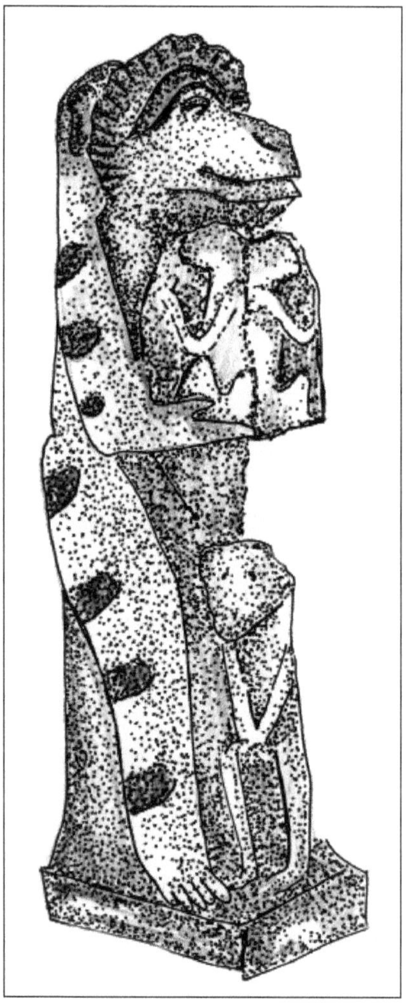

Fig. 49: A monkey amulet in Cairo, JE 31670 (After Bulté 1991: pl. 16, no 16a)

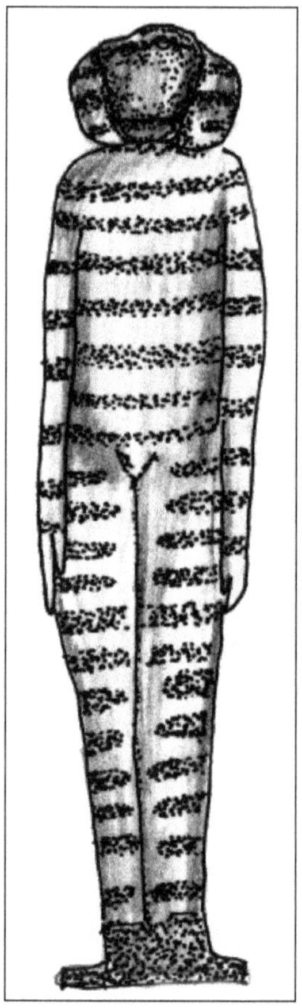

Fig. 50: A walking monkey amulet from the Leo Mildenberg Collection (After Walker 1996: 68, fig. 98)

of a spirit. It often contained a variety of solar images, including the solar baboons. The four sons of Horus were also commonly represented on these amulets, as were Thoth, Amun-Re, Isis, Nephthys, Horus and Re (Ikram and Dodson 1998: 144).

Animal cults had become enormously popular by the Late Period, and most of the known mummified monkeys and baboons date from this period or later. These animals mainly come from cemeteries near the temple complexes of Saqqara and Tuna el-Gebel, which encompasses the western cemetery of Hermopolis Magna. Hundreds of thousands of ibises were also mummified at these sites and deposited at the temples as votive offerings to Thoth. The birds were held by the feet and dipped into vats of molten resin before they were wrapped in linen, but it is unknown whether the procedure was performed while the ibis was still alive. Some of the birds seem to have broken necks, so they may have been killed before being dipped in resin (Ikram and Iskander 2001: 3). A much smaller number of baboons were mummified as votive offerings—only a few hundred have been recovered. The manner in which they were killed has yet to be determined (Ikram and Dodson 1998:136). Some of the baboons, however, were cult animals, believed to have been the incarnation of the god on earth. These would have been allowed to live out their natural lives and were only mummified after death.

In 1964, Emery began excavating the ibis galleries at Saqqara and was struck by the fact that they had been deliberately situated in an area previously occupied by Third Dynasty officials, even though their construction had been hindered by the earlier tomb chambers and shafts. He hypothesized that the galleries may have been built in that area because of their putative proximity to the tomb of Imhotep, the master builder of Djoser's pyramid, who had been deified in the Late Period (Emery 1965: 8; Smith 1974: 27). By the Ptolemaic period, he became conflated with Asclepius, the son of Hermes, with whom Thoth was assimilated. In the temple, at the entrance to the ibis gallery, Emery found ostraca referring to Thoth the Thrice Great, and other areas of the site revealed references to Imhotep (1974: 28). The baboon catacomb is composed of an upper and lower gallery, each containing a series of niches cut in the rock walls to hold the wooden boxes that served as sarcophagi for mummified baboons. The Upper Gallery dates from c.400-200 BCE and contains 200 niches, while the Lower Gallery dates from c.200-30 BCE and contains 237 (Goudsmit and Brandon-Jones 2000: 111).The niches were sealed with a limestone slab inscribed with the name of the animal, its place of origin, the date of burial and a prayer for its survival in the afterlife. Unfortunately, Christians destroyed the catacomb, and only one baboon burial survived intact because it was hidden by a heavy sarcophagus. Surviving limestone panels revealed that mature baboons were installed in the temple of Ptah-under-his moringa-tree[11] as living incarnations of the god. When they died, they became "Osiris the Baboon." One inscription named a baboon as "the Hearing Ear." This title was given to deities that performed an oracular function, and it is probable that the oracular questions were heard by a pair of limestone baboons, whose muzzles had also been broken off by Christians (Smith 1974: 41-2).

At the northern entrance to the tomb of Petosiris, the temple administrator at Hermopolis Magna (Tuna el-Gebel), dating from the reign of Ptolemy I, are scenes involving Osiris-Baboon and Osiris-Ibis. North of this, a temple of Osiris-Baboon had been built in the name of Alexander IV. The Ptolemaic section of the animal galleries (Gallery C) is located at the back of the temple. An annex building, containing a cage for a sacred baboon, apparently in connection with the oracular functions of Osiris-Baboon, was added to the temple's southern side after a violent earthquake (Kessler and Nur el-Din 2002: 36). Excavations into earlier parts of the galleries have also uncovered faience and bronze figurines of Thoth-Ibis, Thoth-Baboon, Osiris-Baboon and Bes, and a hamadryas mummy, dating from the Persian period, was actually found intact in a rock-niche. In the earlier phases, the cemetery contained bones from a variety of animals, but no complete mummies (2002: 37-8).

The discovery of the hamadryas baboon, however, is important for another reason. In the sculpture from North Saqqara, baboons were the most frequently represented animal, and all statues seem to depict the hamadryas (Hastings 1997: 47). During the Ptolemaic period, however, other species were used as substitutes. Goudsmit and Brandon-Jones were able to recover remains of 169 individual monkeys from the catacomb and found that 146 were anubis baboons, 21 were Barbary macaques and 2 were vervets. No hamadryas baboons were found (Goudsmit and Brandon-Jones 2000: 111). Most of the animals died at fairly young ages. Approximately 83% of the monkeys died before the age of thirteen—the average age of baboons is forty, while macaques generally live about thirty years. An analysis of the bones revealed dental and skull abnormalities as a result of a vitamin D deficiency. Monkeys, like humans, absorb D through UV light on the skin, so the animals were probably captured at a young age and kept indoors (Goudsmit and Brandon-Jones 1999: 51). Hamadryas baboons may have been in short supply by the Ptolemaic period, but there is no evidence for the importation of Barbary macaques from western North Africa before this time, even though they were being imported to Etruria by c.450 BCE (Goudsmit and Brandon-Jones 2000:112).

A few monkey mummies post-dating the New Kingdom were still just pets. A monkey found interred with Maetkare, the God's Wife of Amun, was believed to be an illegitimate child that died in infancy, but when it was x-rayed in the 1970's, it was found to be a pet monkey. The simian was first identified as a hamadryas baboon, but now it is believed to be a vervet or patas monkey (Ikram and Iskander 2001: 53; Ikram and Dodson 1998: 329; Brier 1994: 209).

CONCLUSIONS

While Egyptians had a profound respect for monkeys, and particularly baboons, throughout the approximately 1,500 year period of the Old to New Kingdoms, they also saw the humorous side to their behaviour and enjoyed

[11] Thoth was associated with Ptah at Memphis through the tradition of the Memphite cosmogony. Ptah represents the intelligence, Horus the heart and Thoth the will, which is identified with the tongue (Hastings 1997: 47).

depicting the monkey's antics in a variety of media. They also recognized the animal's similarity to humans and imagined the creatures performing almost every human activity, from the sacred to the silly. The Egyptians did not view baboons as ugly, poor imitations of human beings, as many subsequent cultures (including our own) have done, but believed they were superior in some ways because of their perceived piety. They were also to be treated with caution because of their strength and potential for aggression, as personified in the god, Babi. As with other potentially dangerous animals, the power of the baboon was harnessed in spells and amulets for the benefit of the human user. The smaller monkeys were loved for their playfulness as well as the belief that their image could promote human fertility.

Continuity in Egyptian culture is certainly demonstrated by the longevity of these views, but clearly innovations took place that reveal the people's capacity for change. The development of animal cults in the Late Period was, however, an unfortunate degeneration of earlier practices. The dearth of hamadryas baboon mummies dating from the Ptolemaic period suggests that the animals were becoming rare in Egypt, perhaps because too many were sacrificed as votive offerings during the Late Period. Additionally, the deformities caused by a lack of exposure to sunlight indicate that the Egyptians of the period were either unfamiliar or less concerned with the needs of the animal. Despite the fact that many monkeys must have been kept indoors, the hamadryas' connection with the sun was kept alive through vignettes from the Book of the Dead and amulets. The history of Egyptian monkeys, however, does not end in Egypt. As briefly noted, by the Ptolemaic period Thoth came to be identified as the Egyptian Hermes. Because many aspects of the god's personality were transformed by Greco-Roman religion and philosophy, he will resurface in the fourth chapter: Monkeys in the Greco-Roman World. In fact, as I hope to demonstrate in all the remaining chapters, Egypt's monkeys made an impact on many cultures throughout the Mediterranean.

Chapter 2
MONKEYS IN THE NEAR EAST

This chapter is intended as a brief overview of the roles assigned to monkeys in Mesopotamia, the Levant and Anatolia in order to provide a background for the exploration of monkeys in both Minoan and Greek iconography. It reviews and comments upon several articles and one recent thesis written on the significance of monkeys in the Near East. As mentioned in the previous chapter, Egyptian pithemorphic vessels dating from as early as the Old Kingdom were found at a number of sites outside of Egypt, including Byblos on the coast of Lebanon. Clearly, Egyptian simian imagery was introduced to the Levant at an early date, but this was not the region's only source of monkey iconography. Mesopotamian civilisations also exerted a considerable influence on the area, and although monkeys are no more native to Mesopotamia than to the Levant, the animals were represented there from the late fourth millennium onward. One of the most beautiful and best known examples is the gold monkey from the tomb of Mes-Kalam-Dug found by Woolley at the early dynastic Royal Cemetery of Ur (Woolley 1934: 300). Some of these representations were probably the result of contact with the peoples of the Indus Valley, where rhesus macaques were and are indigenous. The Hanuman langur, whose range extends throughout India, may also be depicted. The influence of Egypt in early Mesopotamia, however, is clearly visible in the earliest representations, which were most likely inspired by baboon amulets.

The most recent work on the subject, Azad Hamoto's thesis, *Der Affe in der altorientalischen Kunst*, leaves something to be desired in the interpretation of the iconographical material and demonstrates a certain lack of knowledge concerning the habitats of monkeys. However, it provides an excellent catalogue of approximately 200 representations of monkeys in a variety of media, arranged in chronological order from the Neolithic to the Hellenistic periods. The earliest pieces were amulets found in Byblos, which date from the Neolithic period (Hamoto 1995: 12, nos. 1-3). These crouching stone figures with long muzzles probably depict baboons, the result of contact with Egypt. Seals and amulets made their appearance in Mesopotamia at Ur, Uruk, Tell Brak, Susa and other sites between c.3300-2900 BCE (1995:13-17). It is difficult to know which species are represented on these objects, but some amuletic figurines have the long muzzles of baboons and are seated in the same position as the early representations of Hedjwer. Mallowan identified the simian figurines he excavated at Tell Brak as baboons but acknowledged that an Asian species could also have been depicted (1947: 42). Ratnagar remarks that in a later Larsa period text, a carnelian monkey figurine was mentioned. Since this stone would have been imported from Meluhha, it seems likely that the later figurines do, in fact, depict Asian monkeys (1981: 149). Hamoto maintains that when Asian monkeys appear in Mesopotamian art, they are most likely Hanuman langurs. However, he is under the mistaken impression that macaques live only in China and Japan (1995: 5-6). The rhesus seems a more likely candidate because of its wide distribution and considerable ability to adapt to new environments and foods. Although the Hanuman langur is one of the sturdier and more adaptable colobine monkeys, it is not in the same league as the generalist cercopithecines, such as baboons and macaques. This does not rule out the possibility that the creatures were occasionally imported to Mesopotamia, but their chances of survival in the new environment would not have been as great. Furthermore, Hamoto catalogued two ED III cylinder seals and one figurine from Ur made from lapis lazuli (1995, nos. 30, 31, and 34). This stone would have been imported from the area of modern day Afghanistan, which is also the westernmost region of the rhesus macaque. It seems reasonable to suggest, then, that monkeys were introduced to the Near East from both eastern and western sources throughout its history.

The multiple roles of the monkeys in the Near Eastern cultures may reflect a combination of imported values and local interpretation. Perhaps the most informative study concerning the monkey's significance in Mesopotamia is Sally Dunham's article, "Monkey in the Middle," which appeared in *Zeitschrift für Assyriologie* in 1985. She remarks that two words for monkey existed in Akkadian, *pagû* and *uqūpu*, and one Sumerian word, *ugubi*. The species represented by these words is unknown, and too little evidence remains to identify the Akkadian words as synonyms (1985: 235). One of the contexts in which the creatures are mentioned is as exotic animals, imported as tribute. One of the earliest examples comes from the Sumerian "Curse of Agade," "...monkeys, mighty elephants, 'water buffalos', exotic animals would jostle each other in the public squares" (1985: 236). These animals were probably imported from the east if the translation of water buffalo is correct (1985: 240). Later Middle and Neo-Assyrian kings, including Tiglath-Pileser I, Aššur-bēl-kala, Aššur-nāsir-apli II, Shalmaneser III, and Assurbanipal, mention receiving tribute in the form of monkeys from Mediterranean sites, particularly the Levantine coast or Egypt (1985: 236-9). The monkeys were clearly valued for their exotic nature, and Aššur-nāsir-apli II even reported breeding them (1985: 238).

The words for monkeys, however, could also be used pejoratively. The Guti, in the Sumerian "Curse of Agade," are equated with monkeys (Dunham 1985: 242):

> Gutium are a people who know no inhibitions, with human instincts, but canine intelligence, and monkeys' features—Enlil brought them down from the mountains.

A number of other passages refer to the invading Guti in a similar manner. Ibbi Suen's twenty-third year is referred to as, "[t]he year the stupid monkeys came down from their mountain (or foreign land) to Ibbi-Suen, king of Ur" (1985: 242).

An Old Babylonian scribal text, known as the 'Monkey Letter,' may also comically refer to a man named Monkey, but this is debatable. Cohen maintained that the letter was the result of scribal whimsy. A homesick guard, comically named Ugubi, asks a passer-by to convey a message to his mother. Because the guard is stationed behind the house of the chief musician of Eridu, his name may be a pun. *Ugubi* corresponds to the Akkadian *pagû*, which may have the double meaning of monkey or musical instrument (1976: 271). Cohen translates the text in this way (1976: 272):

> Passer-by! Passer-by! Speak to my mother! Thus says Ugubi:
> "Ur is the joyful city of Nanna!
> Eridu is the prosperous city of Enki!
> I am sitting behind the door of the chief musician's house. I am enjoying this observation post. I shouldn't die from this. The bread isn't spoiled; the beer isn't spoiled."Oh traveller, send this for me. It is urgent.

Powell, however, insists that Ugubi is a real monkey living in the chief musician's quarters and provides a very different translation (1978: 194):

> Monkey to mother:
> Ur is the City-Resplendent of Nanna. In Eridu, the City-Abundant of Enki I am locked up in the house of the Chief Musician. I'm being fed on looks alone. Don't let me die of hunger for fresh bread and fresh beer! Send me some by diplomatic post!
> The situation is desperate!

The humour in the passage then revolves around the business-like style of the letter, and the monkey's request for a diplomatic gift, not of valuable commodities, but of common bread and beer (1978: 194).

Whichever translation is closest to the intended meaning of the passage, monkeys do seem to have been closely linked to musicians and musical instruments. Rutten cites a sealing from the Early Dynastic Royal Cemetery at Ur as evidence of the Mesopotamian use of animals as musicians. In the impression, two highly stylized long-tailed animals, identified as monkeys, sit facing each other on a type of platform. One plays a harp, while the other plays an instrument that resembles a sistrum (**Fig. 51**; 1938: 105). Another cylinder seal from Susa, dating from the ED III period, comprises two registers of images (**Fig. 52**). On the upper register, to the left, a male figure in a short kilt stands on top of a pair of dogs. He carries a bow in his left hand. A goat, seeming to have leapt straight up, looks back at the hunter. The goat's feet are posed just above the head of a lion. This creature is one of an antithetical pair supporting a large warrior goddess, possibly Ishtar, who carries a number of weapons on her shoulders. She looks toward a man, who steps toward her. His left arm is raised, while the right is clasped to his chest. Between them are the images of the sun, a crescent moon and a star. A scorpion with upraised arms floats in mid-air behind the man and to the left of a fantastical winged bird-man. A seated monkey playing the flute hovers just to the right of the creature's legs. To the right of this, a bull-headed man and long-haired male figure flank, and seem to support, a plant (Hamoto 1995, no. 36). The musical monkey, it would seem, was considered worthy of inclusion in a narrative with religious significance.

A series of terracotta plaques from Ur and nearby Diqdiqqah depict a man and one or two monkeys on a leash. The man usually wears a round cap and long skirt. A monkey sits on his shoulder, and he holds the animal's leash in his hand. Some plaques also depict a second monkey sitting at his feet. Although none of the plaques

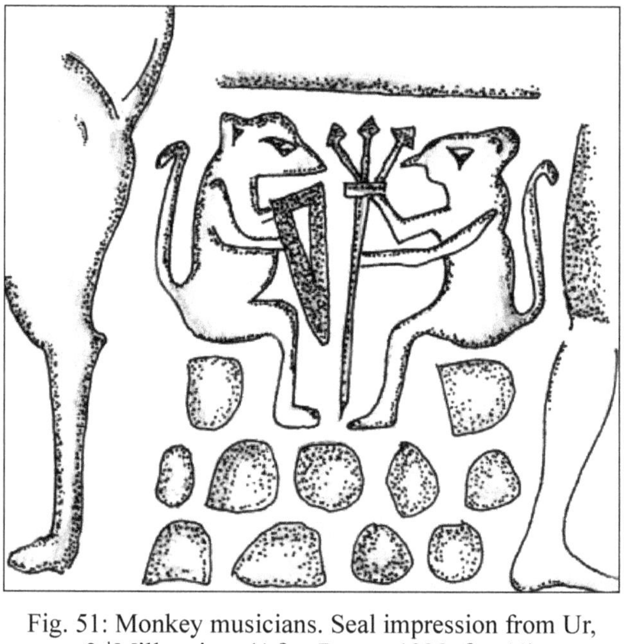

Fig. 51: Monkey musicians. Seal impression from Ur, 3rd Millennium (After Rutten 1938: fig. 11)

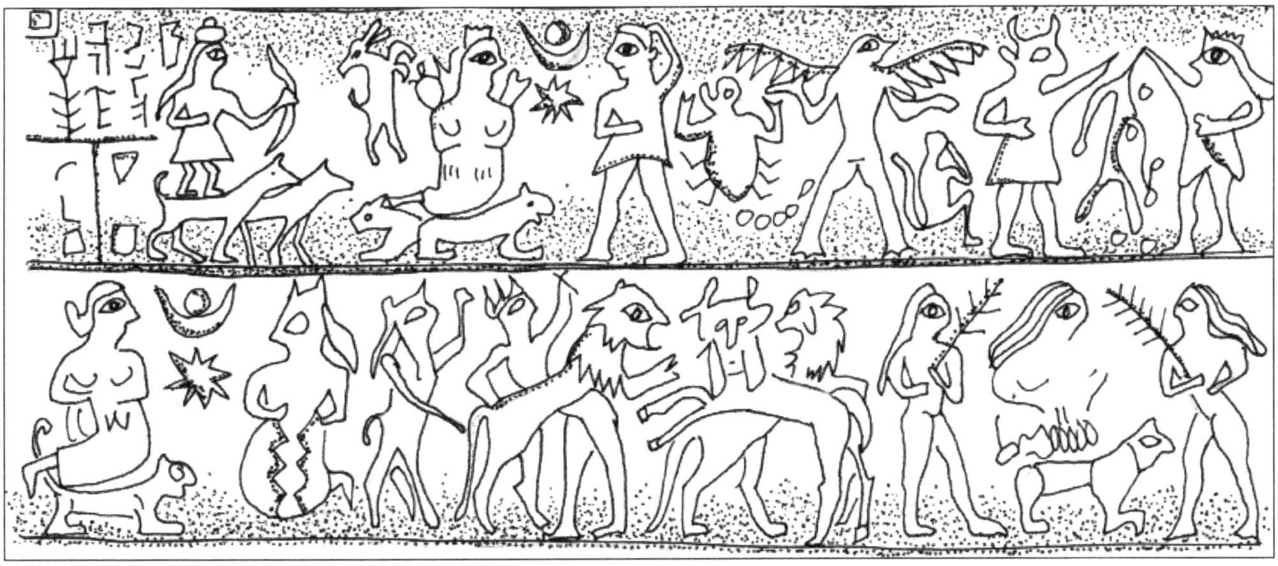

Fig. 52: Cylinder seal from Susa, ED III (After Hamoto 1995: no. 36)

comes from a securely dated context, Dunham dates them to the Old Babylonian period on the basis of the man's attire and the fact that most of the finds from Diqdiqqah suggest an Old Babylonian date. Because plaques from this date often depict professions, she suggests that the man may be a professional monkey handler (1985: 241). Hamoto concurs with the dating but offers no suggestion concerning their significance (1995: 91). Mendleson has suggested an alternative date and interpretation based on a study of the three plaques housed in the British Museum. On the first plaque, a bearded man holds two monkeys by a leash. One of the monkeys sits on his shoulder, and the other sits at his feet playing the flute. The second is broken off at the top, but the man clearly holds a long object, possibly a musical instrument, in his hand. Like the previous example, a flute-playing monkey sits at his feet. In the third example, the man holds an object over his right shoulder that Mendleson describes as a bladder on a stick (**Fig. 53**; 1983: 81). This time, the flute-playing monkey sits on the ground behind him. She suggests that the man is an itinerant entertainer from the Indus Valley, related to the modern-day monkey Wallah of India, and dates the plaques to the Ur III period. The monkey is identified as the common rhesus macaque that is now associated with the Wallah. She believes the plaques are earlier than other scholars have suggested because Mesopotamia's connections with the Indus Valley were greater during the Ur III period. The flute-playing monkeys, however, show the influence of an intermediate culture and may reflect an Elamite fable. Mendleson notes the plaques' similarity to the earlier sealing from Susa, described above, and suggests that a lapis lazuli cylinder seal from the Royal Cemetery at Ur may also reflect the influence of Elam. Additionally, the word *ugubi* first appears in Sumerian literature after the Elamite invasion (1983: 82-3). The Guti or Elamites are described as monkeys descending from their mountain (Dunham 1985: 242; Mendleson 1983: 83).

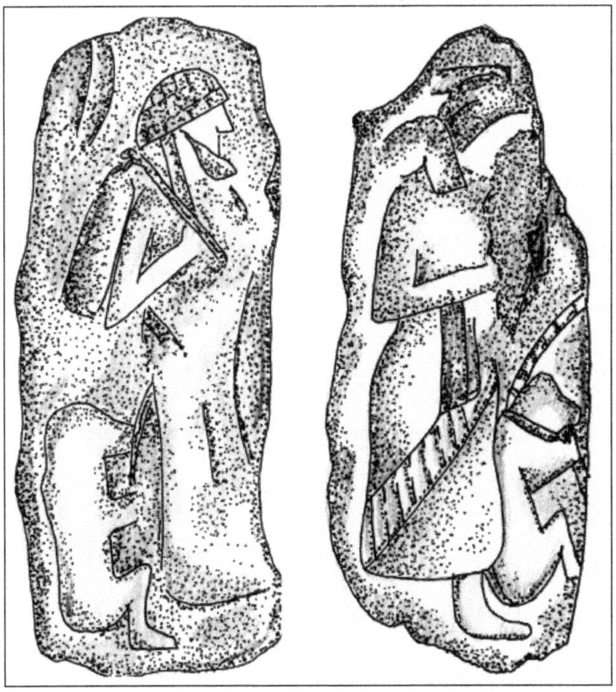

Fig. 53: Two terracotta plaques from Ur, possibly Ur III

Although Mendleson admits that evidence for the Elamite introduction of monkeys is sparse, it is evident that the creature's roles in Near Eastern traditions are the product of cultural interaction. A later Syro-Cappadocian seal from Alalakh, cited by Collon, depicts a nude female, probably a goddess, wearing a horned head-dress, followed by a male figure, possibly a sun-god, who carries a cleaver with a serrated edge. The pair approach an enormous fantastical creature, an eagle with two lion's heads, to the left-hand side of the scene. Between the eagle and the goddess is an open-mouthed dog. A long-tailed monkey sits behind the goddess (**Fig. 54**). Although she does not comment on the monkey's appearance on the seal, Collon remarks that a number of elements from the scene bear similarities to those from seals found at Kültepe, Karahüyük and Ur.

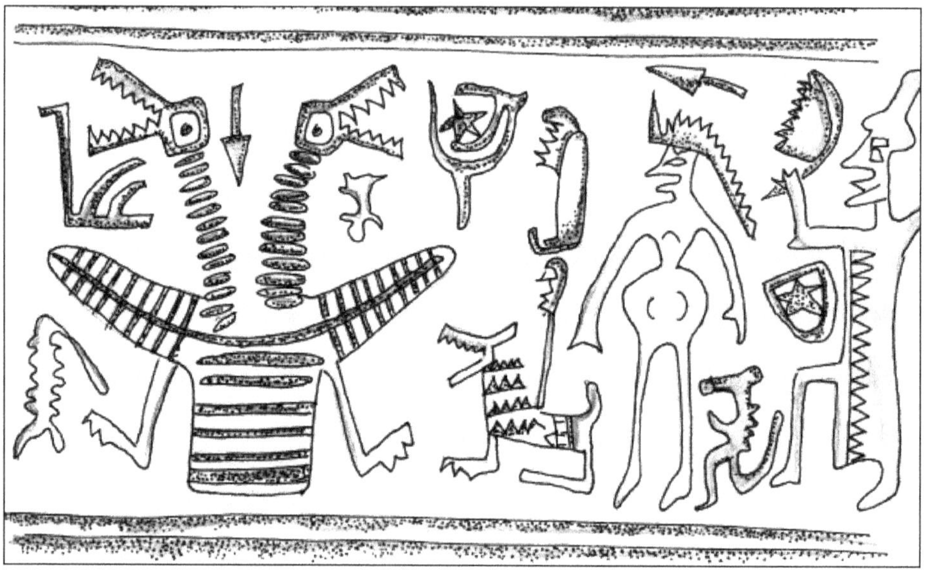

Fig. 54: Black stone seal from Alalack (After Collon 1982: fig. 9)

The lion-headed eagle is apparently an evil monster that is fought by the goddess and god (Collon 1982: 40-41). It seems reasonable to assume that the dog in front of the goddess and the monkey seated between the two deities are present to support their efforts.

A number of other seals combine the monkey's associations with nude women and musical instruments. Dunham cites a round terracotta plaque dating from the Old Babylonian period which depicts two long-legged women walking toward each other, separated by two small bow-legged men, placed one on top of the other, playing lutes (**Fig. 55**). A monkey sits above the men, facing the female figure on the right. Behind her is another seated monkey, facing the central scene. To the left, a long-tailed monkey stands behind the other female and also faces the central scene. This animal appears to be holding a leash or cord and all three are wearing halters around their chests. She suggests that the bow-legged men, monkeys and women acted as performers on certain unidentified occasions (1985: 246, fig. 2). It is tempting to see an Egyptian influence in the plaque, since dwarves were often associated with musical entertainment and monkeys in the slightly earlier Old Kingdom, as were nude female dancers. Indeed, the seated monkeys look very much like baboons.

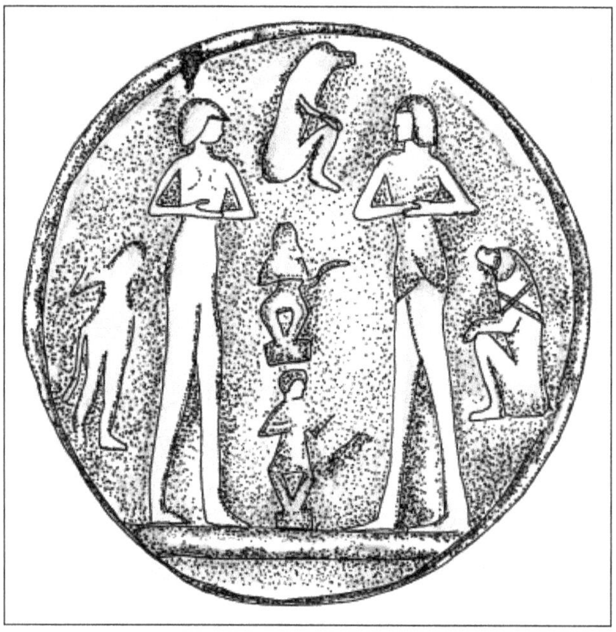

Fig. 55: Terracotta plaque, Old Babylonian (After Dunham 1985: fig. 2)

An Old Syrian style copper cylinder seal at the Ashmolean Museum adds support to the notion that monkeys retained some of their Egyptian associations as they were introduced to the Near East. It depicts a confronted bull and lion sideways on a vertical line, a stylised tree, a kilted man wearing the double crown of Egypt, a monkey seated below a disk and crescent, two confronted male figures separated by a falcon placed below a winged disk with uraei (Buchanan 1966: 177, no. 905). Clearly, traditions concerning Thoth and Horus had enough impact on the Levant for their animal attributes to be recognised. A few Neo-Assyrian ivory figures and plaques demonstrate Egypt's influence as far inland as Nimrud. Two figurines depict a man with distinctly Nubian features guiding a gazelle with his left hand and carrying a monkey on his shoulders. Both figurines are skilfully carved and depict both the men and the animals in a realistic manner. The first monkey is instantly recognisable from its cape as a male hamadryas. The large animal practically reclines along the man's shoulders. Its tail hangs down the back of his arms, and its hindquarter rest directly on his right shoulder. The baboon leans on the man's right shoulder with its left hand and holds something to its lips with its right hand. The second figurine depicts a vervet straddling the man's left shoulder and grasping the top of his head with both hands. It clings to its keeper's torso with both feet. The Nubian bears a characteristic leopard skin over his right arm (**Fig.**

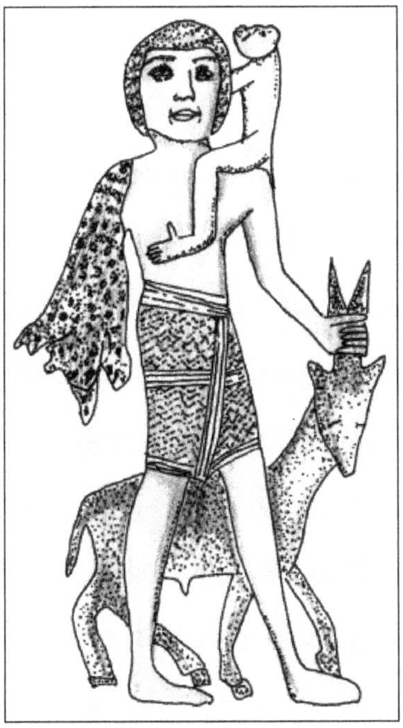

Fig. 56: Ivory figure from Nimrud

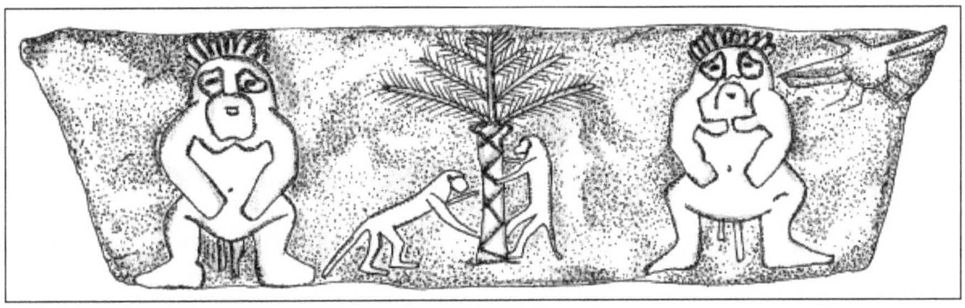

Fig. 57: Ivory plaque from Nimrud (After Hamoto 1995: no. 180)

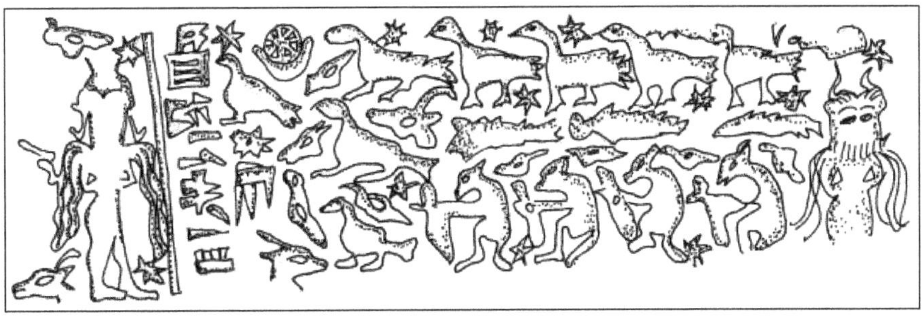

Fig. 58: Anatolian seal with row of monkeys (After Mellink 1987: pl. 19, no. 7)

56; Hamoto 1995: 114-15, nos. 177-78). A plaque from the same find spot presents a highly symmetrical composition of twin images of Bes squatting at opposite sides of the scene. From above, a bird flies toward each image of the god. In the centre, two long-tailed monkeys are climbing a palm tree (**Fig. 57**; Hamoto 1995: 115, no. 180).

Monkeys also had a place in Mesopotamian omens and medicine. Dunham catalogued nine omens from the first millennium Akkadian texts, *Šumma Ālu* and *Šumma Izbu*, concerning other species giving birth to monkeys, walls and body parts resembling monkeys, and dreams of eating monkeys. Although many are incomplete, the majority seem to place the monkey in the context of unfavourable omens. A line from Tablet 88 of the *Šumma Ālu* states, "If a crenel looks like a monkey but when you ascend the wall it is normal, the destruction of Nippur" (1985: 249). An omen of anomalous birth from the *Šumma Izbu* also bodes ill for the city, "If a ewe gives birth to a monkey—an enemy will take the elders of the land; the land will …; the reign will change; the land will go bad" (1985: 249). Dunham suggests such omens echo the "monkey in the mountain" references to eastern invaders (1985: 251). In medical prescriptions from the same period, however, monkey hair and bone were believed to have healing properties.[12] They are mentioned as the components of fumigants and concoctions in leather pouches worn on the body that were used to treat a variety of ailments, including ringing in the ears resulting from an attack of the 'hand-of-the-ghost' and night-terrors brought on by the *bennu* disease (Dunham 1985: 253). It was also cited as a component in a long prescription to remedy impotency (1985: 254).

Further west, Mellink has suggested that the Anatolian monkey's role as a libation pourer was the inspiration for the Minoan genius. While this seems highly improbable, it does seem likely that Anatolian monkeys influenced the development of simian imagery in Minoan culture, as will be discussed in the following chapter. A row of four highly stylized seated monkeys holding libation jugs topped with branches appears on a seal decorated with rows of birds, hybrids and familiar celestial fillers, such as stars, crescents and disks (**Fig. 58**; 1987: 66, pl. 18, no. 7). Mellink remarks that in Mesopotamian art, vases sometimes have plants issuing from the rim to indicate the liquid's importance to vegetation. The monkey's association with life-giving fluids is further demonstrated by its presence in scenes where human attendants make libation offerings to storm gods (1987: 66-7). The monkey's association with libation is visible in an ivory handle from a libation vessel with a carved monkey, dating from the 19th-18th centuries BCE, found at Acemhöyük (**Fig. 59**; Mellink 1987: 67, pl. 19, nos.8; Hamoto 1995: no. 75). A small ivory figurine, probably also from Acemhöyük, depicts a seated monkey clutching a vessel to its chest (**Fig. 60**; Hamoto 1995: no. 73, Mellink 1987: 67, pl. 19, no.9). Mellink describes the monkey's overall function as that of a minor, but magical, attendant of libation rituals that takes the place of human attendants (1987: 67-8). As previously mentioned, pithemorphic vessels were imported to the Levant from Egypt. It seems reasonable to suggest that the Anatolian role of libation bearers was inspired by a combination of pithemorphic vessels and stories of the supernatural qualities attributed to monkeys by the Egyptians.

Finally, a brief discussion of the late depictions of monkeys in scenes of tribute is in order. One of the best known comes from the Northwest Palace of the Neo-Assyrian king, Assurnasirpal II, part of which is now in the British Museum (**Fig. 61**). Tributaries in western clothing bring valuable commodities to the king, including jewellery, vessels and textiles. One brings two monkeys. Winter remarks that the commodities correspond with the

[12] Dunham recognizes that some scholars believe 'monkey bone' and 'monkey hair' may refer to plants but insists that the determinative for plant is not used for monkey bone or hair in any extant examples (1985: 252).

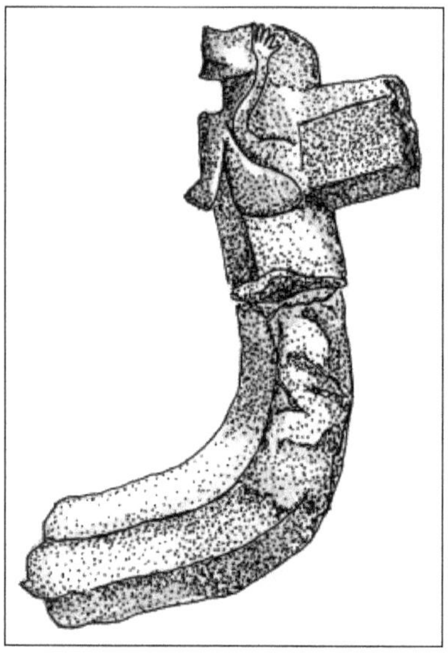

Fig. 59: Ivory handle from Acemhöyük
(After Hamoto 1995: no. 75)

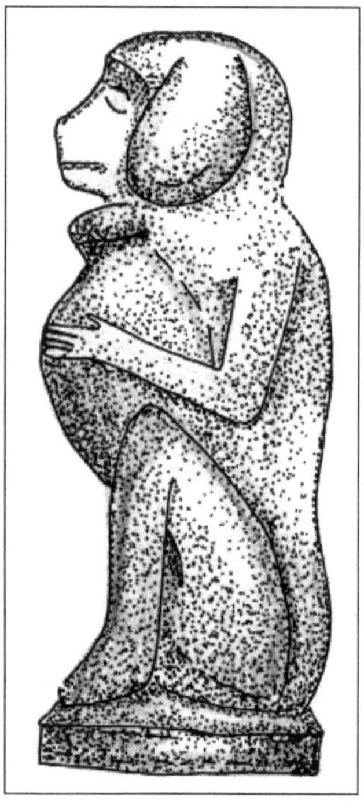

Fig. 60: Ivory figurine from Acemhöyük
(After Hamoto 1995: no. 73)

Ninartu Temple Inscription, which mentions jewellery from Carchemish and two female apes (*sic*) from the Phoenician coast (1983: 27). The two long-tailed monkeys have strangely human faces and hands, but their bodies are simian in appearance. One rides on its keeper's shoulder, resting its one hand on his head and holding its leash with the other. The second monkey walks in front of the keeper on its hind legs, leaning forward on its leash, while looking back with a slightly apprehensive expression. Its overly large hands hang limply toward the ground. Similarly depicted monkeys are shown with tributaries on the earlier Black Obelisk of Shalmaneser III

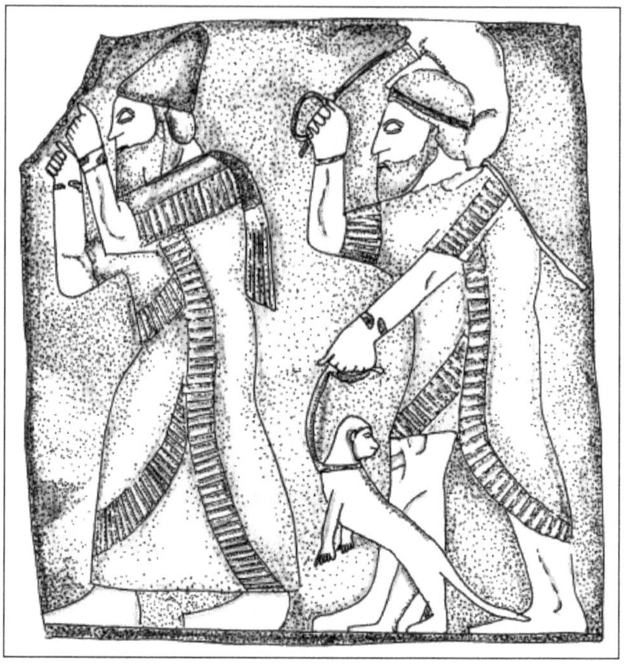

Fig. 61: Tribute bearers from Court D, Northwest Palace
(British Museum 124562)

(**Fig. 62**). On one of the registers, two men with monkeys follow an elephant. The man to the right-hand side of the scene carries a monkey on his shoulder, while holding the leash of another that walks in front of him. This monkey also looks back at its keeper. The man to the left of this monkey holds the leash of another monkey. This animal looks toward the elephant's tail and brushes its leg with its hands. Interestingly, monkeys are mentioned as tribute to Solomon in two Biblical passages, I *Kings*, 10:22 and II *Chronicles* 9:21 (McDermott 1938: 21), and the Black Obelisk also contains a scene of Jehu, a usurper mentioned in *Kings*, paying tribute to Shalmeneser (Collon 1995: 27). Clearly, the Neo-Assyrian period saw the importation of exotic animals, particularly from the West, to all areas of the Near East, and the monkey seems to have been prized mainly for its value as an exotic. The strange depictions of the animals' faces and hands, however, suggest that the Neo-Assyrian artists were acutely aware of the creature's similarity to humans.

Monkeys, then, had a long and varied history in the Near East. It seems reasonable to suggest that, for the most part, they were highly valued as exotics, as attested by the numerous, beautifully rendered figurines made from precious materials. The Sumerian literary references to the Guti, however, suggest that monkeys must have been viewed as somewhat aggressive animals, with a touch of stupidity. A negative view is also apparent in the Akkadian omen texts. Clearly, though, monkeys were never considered entirely without merit. As in Egyptian society, they were valued for their ability to entertain. They also appeared to support deities in their battles against monsters and acted as the attendants of weather gods. In the next chapter, concerning the monkey's roles in the Aegean, many similarities will emerge both in the animal's functions and the materials in which the image was rendered.

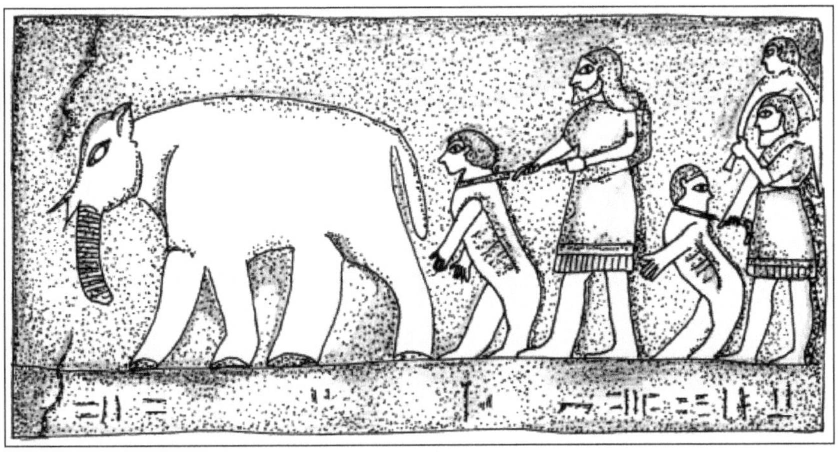

Fig. 62: Scene from Black Obelisk of Shalmeneser III

Chapter 3
MONKEYS IN THE BRONZE AGE AEGEAN

Monkeys, although not native to the Aegean, are well represented in Minoan and Minoanizing iconography. Seals carved in the form of seated monkeys dating from EM II-III have been found in a number of sites from Crete, and by LM IA simians were depicted in at least four frescoes from Akrotiri, two from Knossos, a small fragment from Phylakopi, a number of seals from Crete and a few pieces of jewellery. Many scholars have suggested that the Minoans may have imported the animals to Crete and Thera, and many cite the discovery of a petrified monkey skull on Thera as evidence of this (Galanopoulos 1969: 153-4; Vanschoonwinkel 1990: 36; Hood 1971: 90; Marinatos 1984: 112; Immerwahr 1990: 42; Cline 1991: 40). This alleged skull, however, was first brought to light in a rather fanciful article by Aris Poulianos, "The Discovery of the First Known Victim of Thera's Bronze Age Eruption," which appeared in *Archaeology* in 1972 (229-30). An object, found in 1966 by a supporter of Galanopoulos' belief that Thera was the lost Atlantis, was identified as the skull of a monkey and classified "...among the sub-family Colobinae of the Cercopithecidae family" (1972: 230). The animal's death was then described in graphic detail as it was struck by flying debris and trapped in flowing lava, where it slowly cooked to death. Fortunately, the alleged skull is only a piece of lava, and no monkey died in its formation (Doumas' comment in Sherratt 2000: 735). Indeed, even the X-rays published in Poulianos' article do not resemble a skull.

This does not, of course, rule out the possibility that the Minoans had ever seen the real animal, but Aegean monkeys in wall paintings are never as realistically portrayed as the Egyptian representations. Even in glyptic art, the monkeys are often highly stylized. Obviously, this makes the identification of the species depicted much more difficult than in Egyptian art. Perhaps the best approach to the problem is to consider which monkeys would have been most accessible to Aegean artists, either in nature or art, before attempting to assign a species to individual representations.

The rhesus macaque, whose westernmost habitat is the area corresponding to modern-day Afghanistan, may have been known to the Minoans, who imported lapis lazuli from the same area. Its features, however, do not correspond well with the majority of monkeys depicted in the Aegean wall paintings. This does not exclude the possibility of Mesopotamian simian imagery entering Crete. As discussed in the previous chapter, portrayals of monkeys from that region are often highly stylized, which could have contributed to the stylization of monkeys in Aegean glyptic art. It is worth noting, though, that Aegean paintings generally portray long-tailed monkeys, which would suggest a lack of familiarity with the North African Barbary[13] and point to Egypt as the main inspiration for simian iconography. There, as demonstrated in Chapter 1, baboons and vervets were imported from the southern lands of Nubia and Punt and patas monkeys may have been indigenous. Cultural transference also took place between Aegean and other Near Eastern cultures, and as demonstrated further on in this chapter, a few examples of Aegean simian iconography do have parallels in Near Eastern seals.

The main focus of this chapter is the Aegean adoption of the simian image from Egypt. There is no question of the long history of trade and cultural exchange between the two regions. It has long been argued that the Egyptian goddess, Taweret, inspired the creation of the Minoan genius, and Weingarten (1991) has effectively demonstrated how the image and its significance developed in Minoan culture. The development of this fantastic creature may provide a useful analogy for the adoption of the monkey and help to explain some of the problems Minoan artists encountered in trying to depict an exotic animal. Gill remarks on the representation of the genius (1964:4):

> That no single explanation based on comparison with natural objects could be found to fit all examples is not surprising since the Minoan craftsman was copying not from nature, but from a picture, and that in a foreign artistic style and possibly in a different medium.

When presented with the image of an unfamiliar, exotic animal, or fantastic creature composed of parts from various exotic animals, the Aegean artists often misinterpreted

[13] The Phylakopi fragment only depicts part of a monkey's eye. It would be impossible to speculate on the species.

some of the physical characteristics of the animal in question. Taweret, as discussed in the first chapter, was first depicted on magical wands during the Middle Kingdom. In Egypt, she was portrayed with the head and body of a hippopotamus, who walked upright on her hind legs and had a carapace on her back. The Minoans, who were unfamiliar with the hippopotamus, first depicted the demon with a slightly hippopotamus-like face. However, the face soon developed leonine characteristics, and the genius' feet also evolved to resemble lion's paws. Much later, when the image was adopted into Mycenaean iconography, the genius was depicted with the head of a donkey. Because the artists were working entirely from art, they felt free to substitute the unfamiliar hippopotamus components of the demon with parts from more familiar animals. The Aegean peoples were probably also unaware that the ivory used to produce Taweret's image came from the hippopotamus and that this substance was also believed to possess powers that would reinforce the apotropaic qualities of the image, so a great deal of the original intention was lost.

Like the Minoan genius, Aegean representations of monkeys may also have been based on Egyptian artistic prototypes rather than on the real animal. Because these animals would have been so unfamiliar to Aegean peoples, monkeys seem to have been placed in a category similar to that of the fantastical genius. In his study of the attitudes toward animals in the village of Baan Phraan Muan in northeastern Thailand, Tambiah found a similar phenomenon that may serve as a helpful analogy. The local people group animals into the three categories: domesticated animals, forest animals and deep forest animals, which comprise the elephant, tiger, leopard and bear. The animals of the third category are rarely seen, and they are assigned a special status. The elephant, tiger and lion are used in astrological charts, along with mythical animals, and some symbolize positive characteristics, such as strength and power (Tambiah 1985: 191). Monkeys were probably far rarer to Aegean peoples than deep forest animals are to the villagers of Baan Phraan Muan, so it is easy to understand why they were assigned a special status. Additionally, Aegean artists probably came across Egyptian magical wands with images both of the hippopotamus demon and knife-wielding baboons, which would have added to the primate's mystic. As in Egyptian culture, the monkey seems to have developed very different roles from that of the Taweret demon, but both seem to have had protective functions and both seem to have been endowed with supernatural qualities.

Because Linear A remains undeciphered any attempt to define the animal's roles can only be based on iconographical representations and analogies with its roles in Egypt and, to a lesser extent, in Near Eastern civilizations. Furthermore, the chronological development of the monkey in Minoan culture is not as easy to follow. Despite the early depictions of the monkey on stamp seals, most narrative scenes in glyptic art and wall paintings date from the much shorter period of MM III-LM IA. The issues are further complicated by the fact that the number of extant simian representations from the Bronze Age Aegean is fairly small—almost every example known is considered in this chapter—and controversy exists over the dating of some of the seals. Langdon maintains that no representations of monkeys of Aegean manufacture are known from the period of Mycenaean prominence (1990: 418), but Karageorghis insists that some Mycenaean terracottas do depict monkeys but have been misinterpreted. His evidence, however, is rather scantily based on two objects. One of the putative monkeys, as he acknowledges, is only known from a photograph in the Musée des Beaux Arts, Budapest (1994: 63), while the other is based on the drawing of a fragment of a seated animal discovered in a LH IIIC context in Cyprus, identified as a cat or dog by Tamvaki (1973: 227, no. 62). Only a small portion of the animal's upper body and front legs remain. Although the shape is rather similar to 6th century BCE Boeotian monkey figurines, discussed in Chapter 4, the painted decoration is not consistent with the later examples. This is probably a Mycenaean piece, but it is most likely a dog, as Tamvaki suggests. The Mycenaeans do not seem to have produced simian imagery at all, despite evidence that they were aware of monkeys. A lentoid seal impression for example, found in a LM IIIB context at Chania Kastelli, which depicts two baboon-like monkeys flanking a central object, is probably the work of an earlier Minoan artist. A second impression, attributed to the reverse of the same seal, contains Linear A symbols (**Fig. 63**; CMS V.1. 233). The piece could well have been kept as an heirloom. Cline's (1991) observation that a number of Egyptian monkey figures have been discovered in Mycenaean contexts is considered at the end of the chapter.

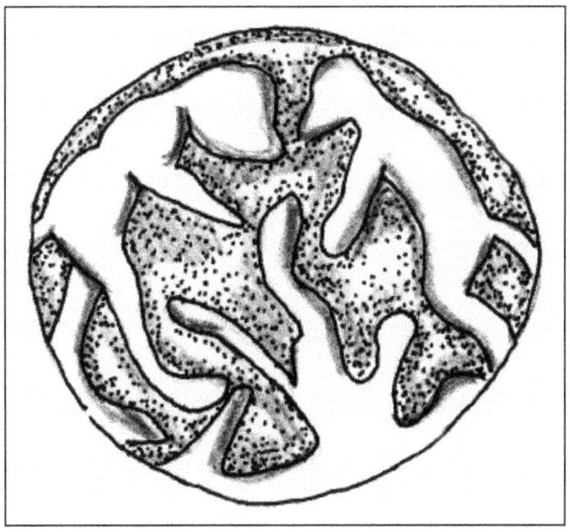

Fig. 63: Seal impression (CMS V.1.233)

The first section of this chapter explores the use and the species of monkeys in glyptic art and categorizes the thematic groupings. The second examines the artistic conventions used to depict monkeys in the Aegean wall paintings and again attempts to identify the species portrayed. This is followed by a discussion of the scene types and analogies between Cretan and Theran representations and scenes in glyptic art. The fourth section explores the possible roles assigned to each species along with some of the cultural and religious associations that may have been adopted into Aegean culture from Egypt along with the monkey's image. Finally, the absence of simian imagery in Mycenaean art is considered. A table at the end of this chapter identifies all the Aegean representations of monkeys I have encountered in my research, including seals, seal impressions, figurines, jewellery, and wall paintings. The dating of the smaller pieces is problematic for the reasons discussed above. Although I have listed the dates in accordance with the CMS catalogues where applicable, I have not found enough evidence to ascribe a date later than LM IA to any of the objects.

SEALS, FIGURINES AND GEMS

One of the earliest Cretan seals in the form of a monkey is the ivory knob seal from Trapeza, Lasithi, dating from EM II-MM I (**Fig. 64**). The monkey was carved in a seated position, supported by its long tail, with its hands resting on its knees. It is one of the more realistic Cretan portrayals, and the monkey can probably be identified as a vervet from its rounded head and long tail. Another ivory seal from Platanos, dating from EM II-MM III, is more stylised and difficult to identify (**Fig. 65**; CMS II.1, 249). It depicts a long-faced, squatting monkey with its hands resting on the ground between its legs. Evans suggests that the lack of tail is an accident of the engraver rather than an indication that the monkey is a tailless Barbary (1924: 119). While I agree that this is probably not a Barbary but a baboon, it seems more probable that the tail was intentionally omitted to preserve the symmetry of the carving. This seal has a number of parallels, notably CMS II.1, 20, from Ayia Triada, CMS V, S.1A, 302, CMS V, S.1A, 303, and CMS X, 30. The posture of the monkeys in this group is highly reminiscent of the Egyptian representations of Hedjwer.

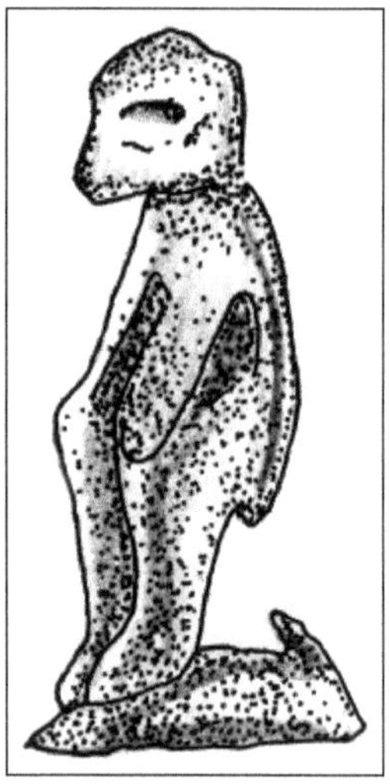

Fig. 64: Seal from Trapeza, Lasithi (CMS.II.1.435)

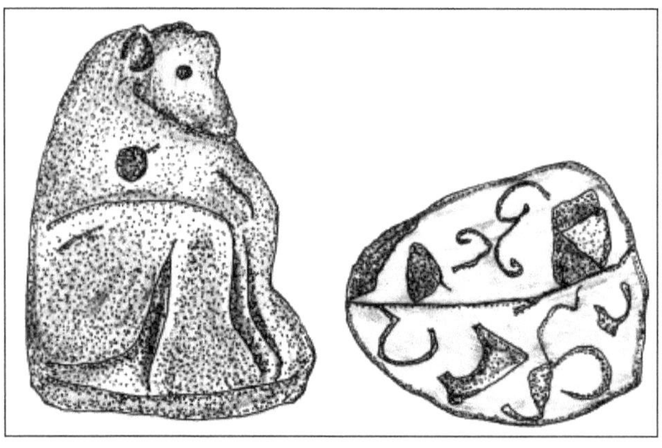

Fig. 65: Ivory seal (CMS II.1. 249)

A few early figurines and pieces of jewellery are also worthy of mention. One is an attractive, highly stylized amethyst monkey, dating from MM II; its find spot is unknown (**Fig. 66**; CMS VIII, 109). Although its face is relatively flat, it has hamadryas-like fur carved around its head. Its ears are covered and the back of its head is large and rounded. The thick fur is emphasized with deep incisions. Its tail is depicted flat against its back in low relief. The monkey rests its left arm on its knee and touches its chin with its right hand.

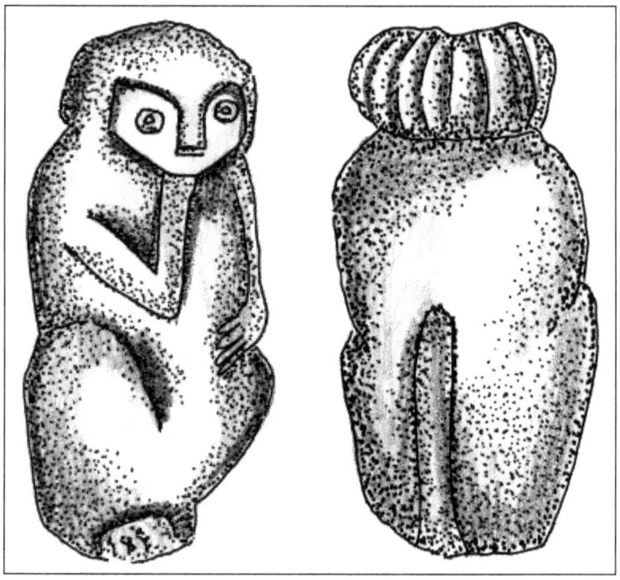

Fig. 66: MM II Amethyst monkey (CMS VIII, 109)

Another monkey carved from rock crystal, dating from MM IB-II, has a similarly flat face (**Fig. 67**; Karetsou 2000, no. 173). This figure, however, seems to be wrapped in a garment. It has a much more rounded appearance than the previous example, and its species is unidentifiable. A lapis lazuli necklace with two monkey pendants, dating approximately the same period, was found in a tomb at Koumasa (**Fig. 68**; Karetsou 2000, no. 174). The monkeys are highly stylized, so, again, the species cannot be identified. Clearly, though, monkeys were considered important enough to represent in precious materials from a very early period. Although Aegean Bronze Age chronology is constantly being redefined, the date for these objects roughly corresponds with the Middle Kingdom in Egypt, where new materials were also being used to create monkey amulets, as discussed in Chapter 1.

Before moving on to a discussion of narrative scenes, two other representations of monkeys in jewellery are worthy of mention. Both have questionable histories, but they are generally believed to date from the LM period, probably LM IA. The first is an electrum ring with a bezel in the form of a crouching monkey from the Aidonia Treasure (**Fig. 69**; Demakopoulou 1996: 17, no. B6). The ring was

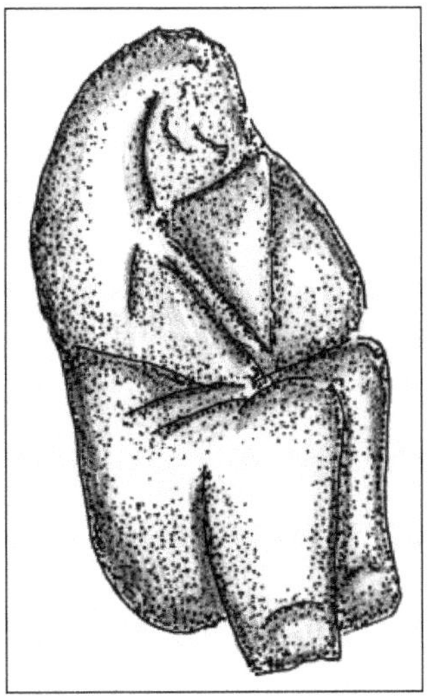

Fig. 67: Rock crystal monkey (Karetsou 2000, no. 173)

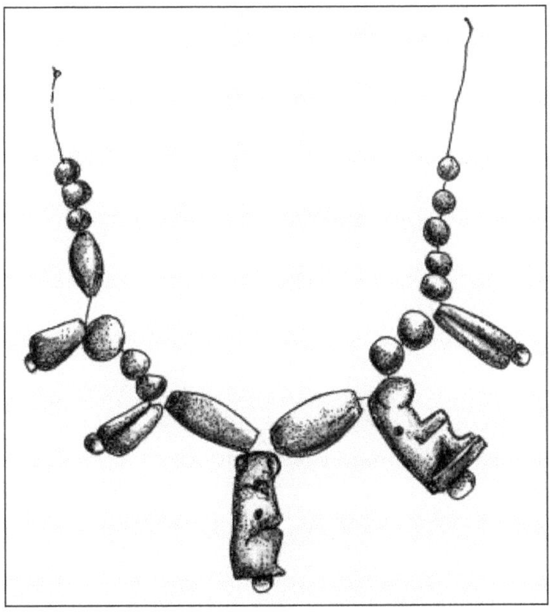

Fig. 68: A lapis lazuli necklace with monkeys (Karetsou 2000: no. 174)

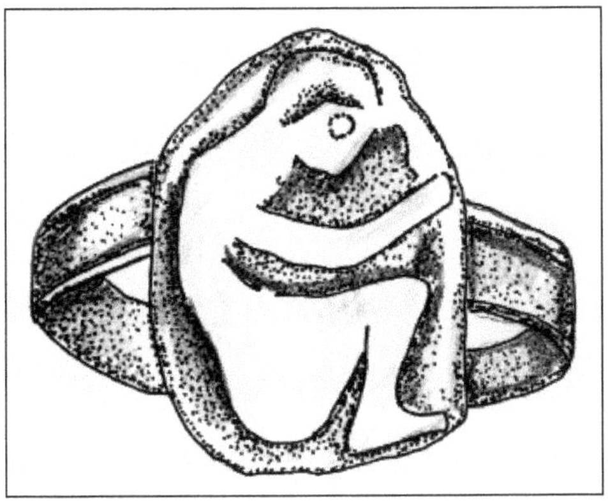

Fig. 69: Electrum ring from the Aidonia Treasure (Athens BE 1996/11.12)

part of an illegal collection of Mycenaean objects believed to have been looted from the cemetery at Aidonia. The monkey ring is thought to be the work of a Minoan artist, which was probably kept as an heirloom (1996: 17). Monkeys are also represented in four gold earrings from the Aegina Treasure, whose interesting and complicated history is traced in a book by Reynold Higgins (1979). Although this collection also contained later Mycenaean objects, the earrings are believed to be the work of Minoan artists. Within a ring in the form of a two-headed snake, a pair of dogs stand on top of two seated monkeys, all facing one another. Attached to the outside of the ring are fourteen chains with alternating ornaments of discs and owls (**Fig. 70**). Hopkins has suggested that the monkeys represent the Egyptian solar baboons and that the dogs are actually jackals, representing the *bau* of the west, associated with the setting sun. Futhermore, the discs may represent the sun and the owls the night (1962: 183). This interpretation is problematic. The baboons are portrayed with their hands held up to their mouths, as if eating, rather than with their arms upraised with the palms facing out in the manner of the *bau* of the east. As discussed below, monkeys are sometimes portrayed with upraised arms in narrative scenes in Aegean art, but the solar element is absent. Additionally, seated baboons and discs in Egyptian art are more likely to have lunar associations, so it is impossible to positively link these monkeys with solar imagery. It is also difficult to positively identify the dogs as jackals; Higgins calls them greyhounds (1979: 26). The role of the monkeys in these earrings, then, must remain unresolved.

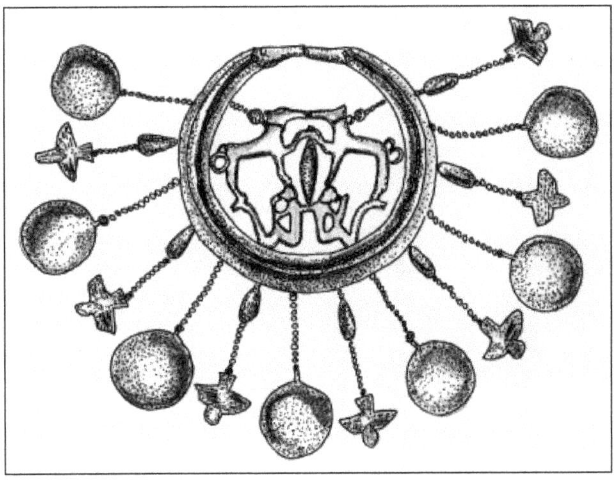

Fig. 70: Gold earring from the Aegina Treasure

Narrative scenes with monkeys, which also began to appear during the MM period, can be categorized according to the attitudes of the animals. Perhaps the earliest narrative scene of monkeys can be found on an ivory signet seal from the context of a MM II tomb at Mochlos depicting two squatting baboons, back to back, with arms upraised in the manner of the Egyptian *bau* of the east (**Fig. 71**; CMS II.1, 473). Evans remarks that the ivory itself is evidence of trade with Egypt (1924: 83). Although ivory could have been obtained from other sources, Aruz has remarked on the strong evidence of Cretan contact with Egypt, which includes the importation of stone bowls, hippopotamus

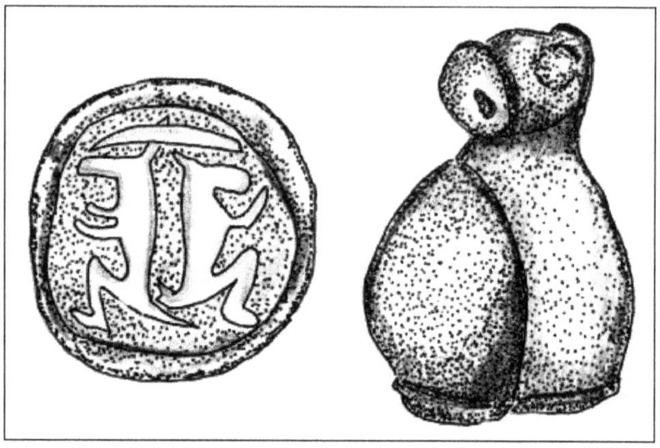

Fig. 71: Ivory seal from Mochlos (CMS II.1, 473)

of the amygdaloid (1958: 61). Marinatos describes the male figure as "imposing" (1987: 127), but the top of the baboon's head actually reaches the man's shoulders—quite an impressive height, considering the baboon is squatting.

Another sealing comes from Zakro, which Evans dates to MM III-LM IA (**Fig. 72**; 1928: 764, fig. 492a). A woman stands to the left-hand side of the impression with her left arm raised and bent at the elbow. Her right arm is bent, with the fist clenched to her breast. The attitude of the woman in the sealing, then, is similar to that of the man in the Giamalakis seal, just described. Facing her is an extremely large, squatting baboon. The spine of the animal has a considerable inward curve, suggesting that it is leaning backward.

and elephant ivory and an ostrich egg (1999: 11). It seems reasonable, therefore, to accept Evans' assessment that the material used in the creation of this seal did come from Egypt. It should be noted further that the depiction of baboons with upraised arms is an indication that the Minoan artist was working from art, not nature, and that it was art of a religious character rather than a scene from daily life. This motif became more common by LM IA.

A later sealing, discovered at Knossos, that depicts a monkey in a similar attitude, was originally cited by Evans, who named it the "Young Minotaur" (Evans 1928: 763, fig. 491). It depicts what Evans thought was a minotaur seated on a campstool, its arms upraised, facing a human figure, whose sex cannot be positively determined because of the fragmentary nature of the impression. Between them is a tree, and the person seems to be pointing toward an animal, which appears to be a ram, resting on the ground, facing the "minotaur." Evans does note, however, that the attitude of the creature is reminiscent of the Egyptian Cynocephalus (baboon). Xénaki-Sakellariou suggests that the animal is a hybrid Cynocephalus-bull (1958: 61). Marinatos' suggestion that the animal is a monkey seems the most probable because of its gesture and the fact that it is wearing a belt (1987: 127). A tree is also present in the scene, which Marinatos perceives to be a palm. She argues that although the palm sometimes appears in Minoan art only as an element of the landscape, it is often represented in a stylized manner with other symbols linking it to cult. These scenes can then be divided into three categories: the triple palm motif, palm with an animal of sacrifice, and palm in ritual scenes with genii, altars and horns of consecration (1984: 115-16). Therefore, the attitude of the monkey and the presence of the palm and the ram would suggest a religious scene.

A seal from the Giamalakis collection depicts another baboon, facing a man, who has his right arm (which is closest to the animal) in an upraised position, bent at the elbow. His left arm is bent, but he seems to be holding it back so that his fingers only reach mid-torso level. He wears a loincloth, and behind him is a series of what appear to be leaves. Xénaki-Sakellariou believes this seal comes from the post-palatial period because of the awkward musculature, short torso, and exceedingly long neck of the human figure, as well as the sharp contours

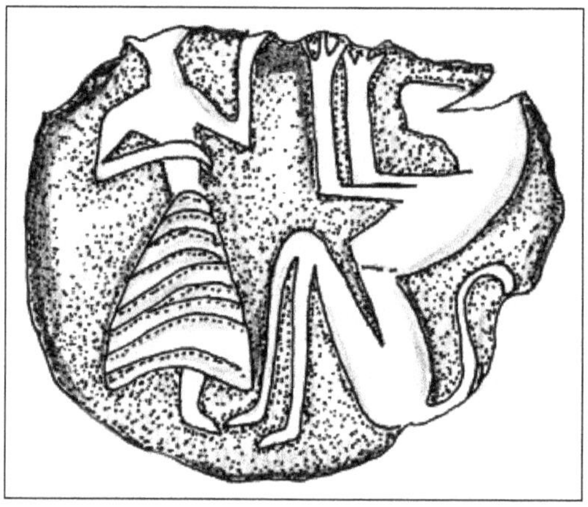

Fig. 72: Seal impression (CMS II.7.24)

On another badly worn seal, a woman appears on the right-hand side of the impression (Giamalakis no. 359). Her arms are pointed downwards, and her left hand seems to be curving upwards, behind her back. Leaves are visible to the far right-hand side of the sealing. Marinatos speculates that because the woman's left knee appears bent, she may have been sitting on a rock, which has not been preserved. If this is the case, the seated position of the woman could suggest that she is a goddess (Marinatos 1987: 125). However, the seal is so badly damaged that I am not convinced that the drawn restoration is accurate. She faces a monkey, again probably a baboon, which may have its head turned away. The animal holds an object, maybe a vase or basket, a detail that makes the scene unique in glyptic art (Xénaki-Sakellariou 1958: 62).

The gold signet ring from the post-palatial cemetery at Kalyvia, Phaestos, probably dating from LM IA, depicts a baboon, standing between two female figures (CMS II.3, 103). The one to the far left-hand side of the scene is wearing the Minoan skirt, whereas the woman on the right appears to be nude and in a seated position, again suggesting that she could be a goddess. Both women have up-raised arms, bent at the elbows. The nude woman, however, is turned so that her chest is shown *en face*. The other woman seems to be turned toward her to allow a ¾ view of her back, and each arm is visible on either side of her body. The baboon is depicted in profile, standing directly before the clothed woman to the left but facing the nude figure to the right, with its arms raised at the elbows. Several other objects are present in the scene, including a pillar behind the nude woman and a plume-like object above the baboon's head, which Evans believed was the stylized ostrich feather used in Egyptian art to represent Maat (1928: 764. fig. 492c). Marinatos reasonably suggests that the pillar substitutes for a tree, introducing another religious element into the scene (1987: 125). It should be noted that, as discussed in Chapter 2, monkeys also appear with nude women or goddesses in a number of Near Eastern seals. The scene depicted on this ring, then, may reflect a Near Eastern influence as well.

The last example of this theme is a sealing from Hagia Triada, representing a lone monkey, sitting in a squatting position surrounded by plants (Evans 1928: fig. 492b). Although it holds its arms in the typical attitude of the Egyptian Sun baboons, the monkey is highly stylized and not immediately recognizable as a baboon.

A second thematic grouping depicts monkeys standing over a central object of importance. The first sealing, discovered at Hagia Triada, dates from LM IA, sometime before c.1450 (**Fig. 73**; Hood 1978: 222, fig. 224b). Two baboons, facing each other, stand over an altar, with their arms pointing straight down, directly over it. Marinatos incorrectly depicts the baboons with male genitalia (1987, fig. 7:1), a feature that is present neither in Levi's drawing nor in the original seal. It is perhaps worth noting here that Aegean monkeys, in striking contrast to their Egyptian prototypes, are never portrayed with genitalia in wall paintings or glyptic art. The scene finds a close parallel in the impression from Chania Kastelli, cited above (**Fig. 63**). Although these baboons are much more crudely rendered, they lean over a central object of importance, with their arms extended downward directly above it.

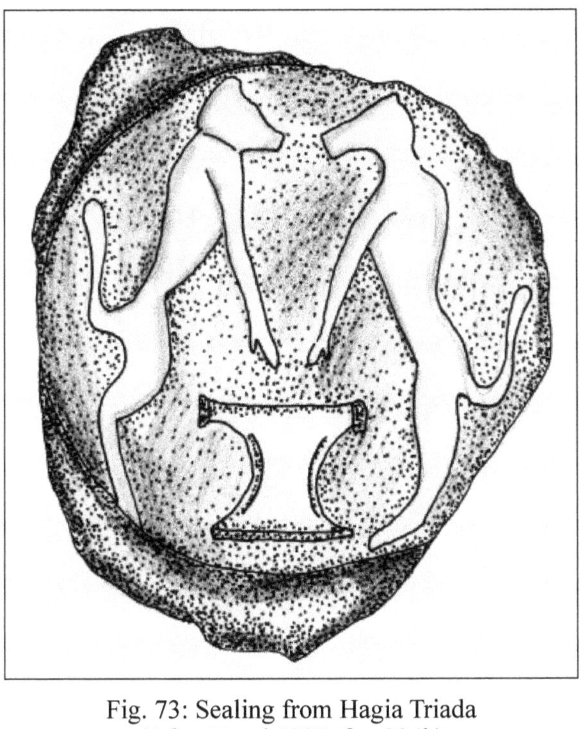

Fig. 73: Sealing from Hagia Triada
(After Hood 1987: fig. 224b)

Another seal with a similar theme, from the Giamalakis collection, depicts two monkeys standing over a vase, which Xénaki-Sakellariou identifies as a kantharos (1958: 56, fig. 355). Their hands meet over the top of the object. It is doubtful that these monkeys are baboons, judging by their relatively short faces and curling tails. Therefore, while they appear in a context similar to the previous sealing, the animals themselves are quite different.

It seems, then, that while the seals in which monkeys appear can be easily grouped into categories according to the attitude of the animals, many different variations occur within the settings. However, certain features, such as plants, perhaps meant to represent a natural element, are common to two or more examples of glyptic art. Religious elements of different forms can also be found in many of the scenes, namely the libation vessel and altar in the last two scenes, the tree in the "Young Minotaur" sealing, and the pillar in the Kalyvia ring. As Xénaki-Sakellariou states, columns and altars often appear in heraldic scenes, flanked by animals, and kantharoi are used in libation rituals on talismanic gems (1958: 56). The evidence suggests that religious and/or natural images do occur with monkeys in glyptic art. It is also noteworthy that most of the monkeys portrayed are baboons. Clearly, the notion that the animals possess religious significance was accepted into Minoan culture, and some of their attributes were retained. They appear displaying the same gesture of respect as the solar baboons in Egyptian art and interacting with figures that probably represent deities. The exact nature of their religious functions, however, cannot be deduced from the glyptic art alone, but the presence of different species in similar scenes indicates that the Minoans did not adhere to the Egyptian categorization of species. This tendency to mix the functions of various species becomes even more apparent when the wall paintings are considered.

CONVENTIONS AND SPECIES IN WALL PAINTINGS

In the LM IA settlement of Akrotiri, monkeys are shown in four separate areas: Beta 6; Xeste 3, room 3a; Xeste 3, room 4; and in Sector A, in the North Magazines. These, like all Aegean monkeys, are painted blue. Vanschoonwinkel remarks that this is departure from the Egyptian convention, in which monkeys are usually green (1990: 336). In many Egyptian paintings, though, animals' colours are realistically portrayed. The fur of the male hamadryas is often portrayed as a light grey, while that of olive baboons and female green monkeys is brown with a green tinge, as the colours appear in nature. Vignettes from the Book of the Dead, on the other hand, often depict the hamadryas' cape or all of its fur as green, and in faience figurines produced before the New Kingdom, a pale blue-green colour was the only option. By the 18th Dynasty, however, many colours were available. One scholar also explains that the Egyptian blue used as paint often experiences chemical transformations when exposed to unstable environments. The blue may turn black or green (Green 2001: 44). In this case, some of the green baboons may originally have been blue.

As noted in Chapter 1, the lunar baboon figurines were usually depicted in blue. A few solar baboon amulets are also blue but seem to be a lighter shade. This may reflect their association with the day sky rather than the night sky. Since these figurines and amulets were so common in the New Kingdom, it seems probable that Minoan artists adopted the colour blue for all monkeys. In fact, the colour itself has been identified as Egyptian blue (Evely 1999: 144-45). A second trait that Aegean monkeys have in common is the white stomach, cheeks and brow-ridge. It should be noted that the conventional blue and white colour scheme does not allow for a distinction between the sexes in species where males and females have distinctive colouring. This combined with the absence of genitalia suggests that Aegean monkeys were intentionally depicted as genderless—a clear departure from the Egyptian portrayals.

Apart from the invariable colour scheme, certain minor characteristics emerge in the individual frescoes to make the sets of monkeys unique. Vanschoonwinkel makes the observation that the eyes of the monkeys in Beta 6 are round, whereas those in Xeste 3, room 4 have almond shaped eyes (1990: 332). This may be true of the Xeste 3, room 3a monkey as well, but much of the face is missing. The latter does have an exceptionally elongated face, with a semi-circular nostril painted at the end of its nose. Those in Xeste 3, room 4 have a strip of blue running along their faces, beginning at the white brow ridge and terminating just before a brown/black patch at the end of their noses. All Xeste 3 monkeys appear shaggier than the Beta animals because of the jagged lines painted along the blue and white borders of their thighs and stomachs. The Beta 6 monkeys have black faces down to their upper lips, which have black dots on them.

With these few differences noted, one can proceed to identify the species represented, beginning with those in Beta 6 (**Fig. 74**). These monkeys can reasonably be identified as vervets because of their black faces, but this is not an exact match. A vervet's facial skin is entirely black, whereas the faces of the Beta 6 monkeys are black only down to their upper lips. However, the lips have been spotted with black dots, most likely representing whiskers. Since this is the only facial hair *Cercopithecinae* have,[14] it seems possible that the artists were presented with the difficulty of rendering a monkey that had a bare face surrounded by fluffy white hair. The result leads to a slightly unrealistic depiction. Doumas describes the tails of the Beta 6 monkeys as prehensile, which is inaccurate, since no Old World monkeys have prehensile tails. However, the comment helpfully draws attention to the animals' long tails. The vervet probably has the most useful tail of all Old World primates.[15] It is also important to recognize that the simians in this painting are highly anthropomorphic, as are all Theran representations of monkeys. The curve of the back, shape of the hips, and long, tapering legs are all very humanoid. The animals look more like graceful, human acrobats than monkeys clambering over rocks; their bodies might even be compared with those of the Bull Leapers from Knossos and Avaris.

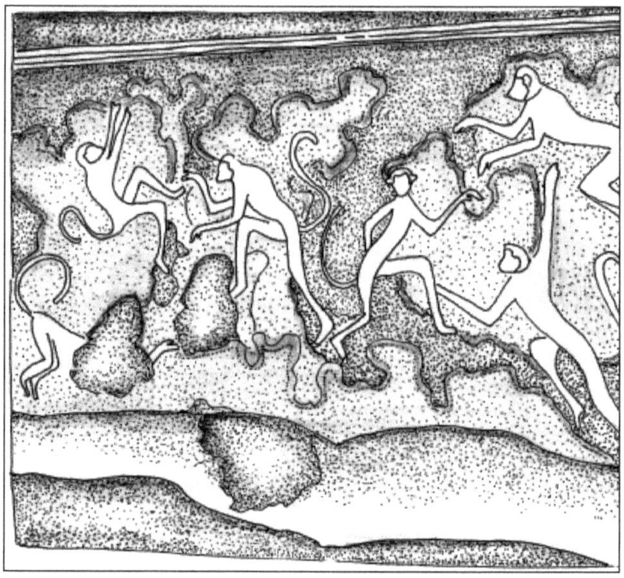

Fig. 74: Blue monkeys in a rocky landscape from Beta 6

The two Xeste 3 frescoes can be considered together because of their striking artistic similarities, even though two different species may be depicted. Their elongated muzzles are more reminiscent of baboons than vervets, but the long tails suggest a hybridization of the two by the artist. One problem with the identification of the Xeste 3, room 4 monkeys as baboons, however, is the dark stripe running along their muzzles, terminating in a black/brown spot. In this respect, the animals resemble the patas monkey. Perhaps this species has never been suggested because the patas is most striking for its red fur, but the facial patterns match this species well. The Xeste 3, room 3a monkey, on the other hand, is most likely a hamadryas baboon **(Fig. 75).** Although its face is badly damaged, it is a peachy flesh colour, devoid of hair and identifying markings.[16] Like the Beta 6 monkeys, the baboon has the rounded back and hips of human figures. Birtacha and Zacharioudakis have argued that the Theran artists made use of stereotypes (templates) to produce such curves. The curve of the monkey's back and hips seem to match those of a number of female figures in the surrounding painting. The tail even falls along the same curve as their skirts (2000: 161-163, fig. 5).

Little can be said of the final fresco from Sector A because of its poor condition. One unusual feature of this monkey, though, is the strangely deformed arms, which seem to terminate in club-like wrists. This contrasts sharply with the well-formed, slender arms and wrists of the other monkeys. Ignoring this fact, however, one can see that the animal has the shorter face and colouring of the Beta 6 monkeys, making it a vervet-human composite.

At Knossos, two frescoes have been found. One is the famous Saffron Gatherer from the Early Keep, which Evans mistakenly restored as a blue boy. Possibly dating from MM IIIA, this may be the earliest extant fresco from

[14] As explained in the introduction, this is a sub-family, which includes the genera of baboons, patas monkeys and guenons, including vervets and grivets.

[15] A young vervet is able to wrap its tail around its mother's for support when riding on her stomach. This makes the animal's tail unique and aids in its identification (Novak 1991: 461).

[16] Because the artists were able to portray these details with such accuracy, it seems likely that they had seen real monkeys. This does not necessarily mean that monkeys were imported—the artists could have spent time in Avaris.

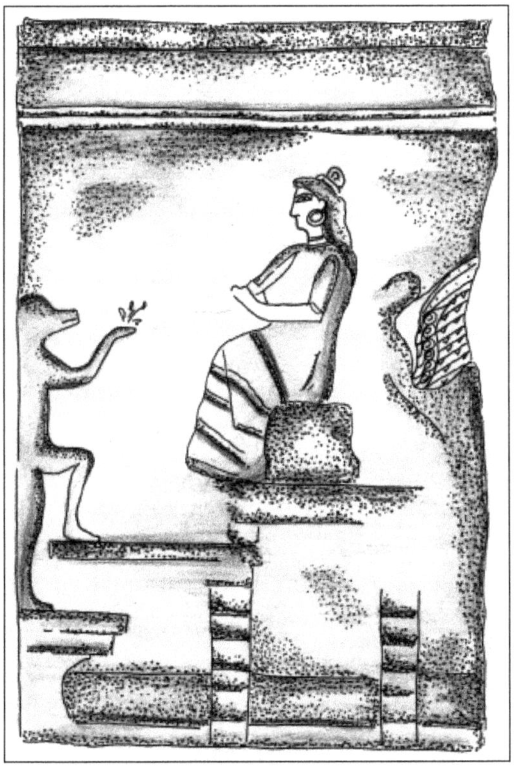

Fig. 75: A baboon offers saffron to a seated female figure from Xeste 3, room 3a

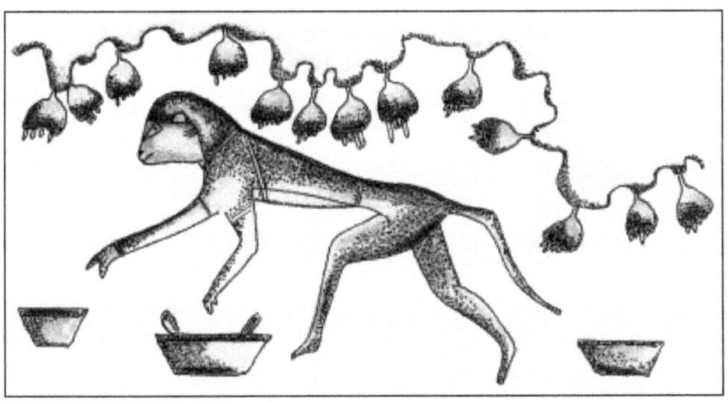

Fig. 76: Saffron Gatherer from the Early Keep at Knossos

Crete (**Fig 76**; Hood, 1978, 48-9). Evans' error, as Hood suggests, probably came about from a misinterpretation of the monkey's belt and armbands as human clothing. The mistake is understandable; however, as noted in the previous chapter, pet monkeys are commonly portrayed in Egyptian art wearing belts, bracelets, collars and anklets, without any attempt to anthropomorphize the animal. This seems to be the artistic prototype of the Saffron Gatherer, whose proportions are entirely simian. The surviving fragment of the monkey facing the Saffron Gatherer reveals that the elongated muzzle is a peachy flesh colour, which is very different from the black-faced Beta 6 monkeys. Since the muzzle seems to be rounded, and the monkey's body is too slender to suggest that a large species, such as baboon, is represented, it seems possible that this is a macaque. As mentioned in the introduction, the two species that would have been available in the Mediterranean were the rhesus from the area of modern-day Afghanistan and the Barbary from Western North Africa. Either species is possible, but it may be worth remembering here that the lapis lazuli necklace, HM Y 146, with seated monkey figures was found at Knossos, which may suggest some knowledge of the rhesus macaque. Of course, lapis lazuli was also imported to Egypt, where it had significant religious connotations, so its link to rhesus macaques is tenuous at best.

The monkeys from the House of Frescoes are also entirely simian in appearance. Indeed, no trace of domestication is present. The monkeys are portrayed running free, unrestrained by belts or collars and unadorned. Some of these have well-preserved faces, revealing a reddish area around the eyes, running down toward the nose, underlined by a dark stripe above the fluffy white cheeks. Since the monkeys seem to have long tails, it seems probable that one of the smaller species of monkeys, possibly an artistic hybrid, is represented.

Before continuing to the next section, more should be said about the Theran choice of representing monkeys with anthropomorphic bodies. It would seem that the Therans viewed monkeys as the only real animal that could take on human characteristics—all other animals are naturalistically represented. Although their exact views on monkeys cannot be expressed with any certainty, it may be helpful to return to Tambiah's analysis of animals in Baan Phraan Muan once more for an analogy because monkeys there are assigned the unique status as human descendants. Recognizing a greater similarity between humans and monkeys than any other animal, the local people created the myth that monkeys were once the human children of a woman who was unable to provide for them. They had to go to the forest in search of fruit, where they eventually developed the characteristics that transformed them into wild simians (Tambiah 1985: 191). This is not to suggest that the Therans had the same story, but it seems probable that Therans also recognized monkeys as the animals most similar to them. The fact that this anthropomorphism is absent in the Cretan paintings suggests at least some difference exists in the cultural attitude toward the animals. The Theran representations would also indicate that a number of narrative traditions surrounded monkeys.

SCENE TYPES

Now that the species have been identified, the scenes in which the monkeys appear can be described and categorised. Each of the Theran paintings depicts monkeys in a different scene type, and each is in a category of its own. The Knossian frescoes both find analogues in those from Akrotiri and will, for that reason, be considered with their Theran counterparts. Then any correspondence between the painted scenes and glyptic can be considered.

The first scene to be discussed is the very fragmentary piece from Sector Alpha in the North Magazine. The vervet squats with its arms up-raised in the manner of the Egyptian Sun Greeters and faces an architectural structure that is religious in nature, either an altar or a hill-top shrine (Doumas 1992: 184; Marinatos 1984: 112; Morgan 1988: 22). The left-hand side of the structure can be seen in the extant pieces, and it is clearly supported by a column, the capital of which is decorated with papyri. Directly above the capital is the end of a flat slab topped with horns of consecration. Little else can be seen in the painting, except that one more monkey was present, since another blue tail is visible to the right of the column. Unfortunately, this scene must be considered out of context, since only equally fragmentary sections of the other frescoes in the area have survived, such as the one depicting the head of an African man, facing a palm tree (Doumas 1992: 187). Obviously, connections between such scenes cannot be made, unless other fragments are discovered. Therefore, all that can be said is that a religious element is present in the scene, as well as some Egyptian influences in the papyrus capital and in the attitude of the monkey.

The next scene type that occurs in the paintings comes from the upper story of Beta 6 and covers the west and north walls (Doumas 1992: 110). Eight monkeys jump and swing through a rocky, stylized landscape, and a river runs along the bottom of the fresco. The two monkeys on the west wall are leaping forward, their arms stretching out as if to catch hold of something when they land. The one toward the top of the scene is looking over its shoulder. On the north wall, the first monkey, to the far left-hand side, seems to be leaning downward, either about to spring up or having just landed. Directly above this animal, another simian is shown with its arms stretched straight up, hands clasped, clearly swinging from something overhead. Another monkey faces these two and has its right leg raised in the manner of climbing or running in an upright position. The fifth monkey, to the right of this one, seems to be running also and is shown *en face*. Only small fragments exist of the last two, but they have been reconstructed in the attitudes of swinging by one arm and jumping.

These are the only narrative paintings from Beta 6 that survived the seasonal torrents that destroyed much of Building Beta, but fragments of what may be part of the continuous composition, including plants and animals, were also found. Sp. Marinatos, the original excavator, listed them as "[a] head, fragmentary of course, may be that of a dog or bovine. Swallows fly in the air. One of them is entirely preserved...great fresco fragments were already found in 1968 in the same neighbourhood showing floral motifs of myrtle and rush" (1970: 63-4). This led N. Marinatos to create a reconstruction of the adjacent wall with two goats or cows, a swallow and the myrtle and rush plants (1984: 92). Of course, such a detailed restoration is speculative at best, but the fragments do indicate that the landscape scene was not meant to be so barren. Therefore, the depiction of monkeys in landscape scenes, where plants and swallows may also have been present, indicates a correspondence between this fresco and that from the House of Frescoes at Knossos.

The Monkeys and Blue Birds painting from the House of Frescoes, dating from MM IIIB/LM IA, was found in an unusual context. When first discovered in room E of the building, which seems to have been a simple house, the fragments, thirty-four in all, were lying in a large but neatly stacked pile. Eventually, eighty-four trays were filled with the fragments (Graham 1987: 58). Evidence has been found for at least six monkeys and eight birds, identified by Morgan as rock doves (1988: 66). Waterfalls and streams run through the rocky terrain, and Marinatos has remarked that many different types of plants that, in reality, could not exist in the same environment are depicted together. Papyri and rush, marshland plants, are shown alongside crocuses and lilies, which normally thrive in rocky areas (1984: 92). As previously noted, the humanoid features present in the morphology of the Beta 6 monkeys are absent in these simians. Even the attitudes of the creatures are more suited to those of real monkeys.

Eggs appear to be present in the published fragments, and it is generally accepted that the monkeys are engaged in an egg hunt—a common activity for monkeys in the wild, despite erroneous comments that only captive monkeys eat eggs (Morgan, citing Cameron 1988: 39). Cameron even restored a monkey eating one, although this appears to be based on a very fragmentary section of the monkey's head. Another monkey, restored by Gilliéron, is pulling back a papyrus stalk, perhaps with the intention of raiding a nest. The awkward angle of the monkey's arm and shoulder reveal the artist's difficulty in portraying the movement. This complicated natural activity is a contrast with the Beta 6 monkeys, who merely seem to be playing. However, both sets of animals are depicted in rocky landscapes with water courses, and the fragments of myrtle and rush may demonstrate further similarities between the two landscapes.

Next are the scenes from Xeste 3, room 4. Doumas notes that this room was connected to three other ground floor rooms through a series of pier and door partitions and that this limited the space available for wall paintings to a narrow strip above the doors (1992: 128). Among the extant images are swallows feeding their fledglings and monkeys engaged in human activities. In one fragment, a monkey plays a lyre or harp, and another depicts at least two monkeys engaged in a sword fight or dance. A hand holding a sword is visible in the left-hand side of the fragment, and Doumas believes that this has been pulled out of a scabbard held by another arm (1992: 95-6; 1985: 31). Facing the sword is another monkey, which seems to have one knee raised and its arms held tightly together, as if grasping something in its hands. This might be a sword that does not exist in the surviving fragments. As discussed in the previous chapter, such depictions of monkeys engaged in human activities, particularly as musicians and dancers, find many parallels in Egypt.

Now the final monkey fresco from Akrotiri, Xeste 3, room 3a, can be considered, followed by a comparison with the Saffron Gatherer from Knossos. Despite some controversy over the exact nature of this room, it is generally agreed to have had religious functions.[17] Five steps lead down to the room from the ground floor, and, on the ground floor level of room 3a, above this room, on the north and east walls, are paintings of a religious nature. The east wall depicts an altar topped with horns of consecration. Red droplets on the horns may represent blood.

On the north wall, three female figures are depicted in a rocky landscape with crocuses. The one to the far left is

[17] Marinatos insists that the room is an *adyton*, and not a lustral basin, on the grounds that the floor would not have been waterproof and no drains are attached to remove the water used in ceremonial purification (1984: 114).

holding a necklace in her left hand and walking toward the altar on the east wall. The central figure, again facing the altar, sits on a rock and holds her right foot, which is bleeding. She holds her left hand to her head in a gesture of pain. A crocus lies at her feet. The easternmost girl has her back to the altar but looks over her shoulder toward it.

Wall paintings are also present on the upper story of room 3a, again on the east and north walls, forming a continuous scene of girls engaged in gathering saffron. This scene almost definitely relates a narrative, which begins with the girl to the right-hand side of the east wall. A basket rests on a small hill behind her, while she partially climbs another small, rocky hill to collect crocus flowers, a cluster of which she holds in her left hand. On the other side of the hill stands another girl, holding a basket in her left hand. This one seems to be absent-mindedly plucking at another cluster of crocuses with her right hand, while conversing with the first girl. This one is clearly older than the first, who has a completely shaved head, except for one lock of hair looped on the back of her head and a tiny front lock, visible over a band on her forehead. Davis maintains that the hairstyle portrayed here represents the first strands of hair that a child is allowed to grow. The girl with whom she is conversing is in the third stage of development. She has stopped shaving her head, and the long side locks she would have grown between these two stages have been cut off. Short, curly hair has grown in and is held back by a small band, similar to that worn by the first girl. The fact that she is older may also be represented by her double chin. However, the blue lines in the whites of her eyes suggest that she is still young (1986: 399-401). This difference in age leads Doumas to suggest that the younger girl is seeking assurance from her elder that she is doing the job well (1992: 130).

On the north wall, a third girl, who appears to be about the same age as the second, judging by her hairstyle and the blue lines in her eyes, carries a basket of flowers on her left shoulder. She seems to be looking downwards, oblivious to the spectacle in front of her: a woman seated on a tripartite platform, flanked by a monkey and a griffin. In fact, only one girl, again in the third stage of development, seems to be aware of this scene. She looks the seated woman in the eye, while emptying her basket into a large container. Thus, almost every step of the crocus gathering process is demonstrated in the scene. The separation of the flowers from the stigmas and stamens, however, is omitted.

The woman sits on the central and most elevated level of the platform. Her height above the other figures is increased by the fact that she is seated on a stool or stack of cushions. On the lower platform behind her, a griffin, which seems to be tied by a lead, stands on its hind legs, wings outstretched, facing her. Marinatos rightly argues that the combination of these factors signals the woman's divinity (1987: 124). Her connection to saffron is also apparent from the depiction of the spice on her cheek and dress, the border of which is also decorated with crocus flowers. She is also connected to marshland through the depiction of ducks and dragonflies attached to her necklaces. A highly anthropomorphic baboon-like monkey stands facing the goddess, with one of its feet on the same level as the saffron gatherers and the first platform. It holds a bouquet of saffron, the finished product, in both hands. The goddess' fingertips are visible directly above the saffron as she reaches to take it with her right hand.

The monkey's connection to saffron is also apparent in the second scene from Knossos, the Saffron Gatherer. In this extremely fragmentary piece, at least two monkeys are present and seem to be engaged in plucking the stigmas from crocuses. Although no human figures are present in the extant fragments, the fact that the monkey is wearing a belt and the crocuses are growing from pots clearly places the scene in a man-made setting.

Now, the similarities between narrative elements in seals and wall paintings can be considered. The attitude of the monkey from Sector Alpha corresponds well with the first category of scenes in glyptic art, and the altar supported by a column with papyrus capitals can perhaps be seen in an abbreviated form in the seals. A pillar, which is probably meant to represent or substitute for an altar, is present in the Kalyvia ring. The tree in the "Young Minotaur" may serve the same function.

The landscape scenes do not fit directly into either category of glyptic art, but the presence of plants does find a certain correspondence in several of the seals. The Giamalakis seal nos. 372 and 359, the Hagia Triada seal and the possibly the Kalyvia ring all have representations of leafy plants in a stylized form.

Finally, the Xeste 3, room 3a painting of the goddess flanked by a monkey and a griffin is remarkably similar to the Giamalakis seal, no. 359. As mentioned above, the seated position of the woman in the seal suggests that she is also a goddess, and the monkey, although squatting rather than standing, seems to be handing her a basket or a vase. The heraldic element of the painting is also present in seals of the second category, in which two monkeys flank an object of importance.

ROLES AND ADAPTATIONS FROM EGYPTIAN ART

In the first section on glyptic art, I argued that Egyptian influence is apparent even in the earliest Minoan depictions of monkeys and that a high percentage of the simians depicted could clearly be identified as baboons. Their gestures and the scenes in which they appear suggest that some of the animal's religious associations were also adopted by the Minoans and the Minoan-influenced Therans. The following section revealed that certain artistic conventions used in wall paintings were adopted from Egypt, while others, such as the depiction of gender, were deliberately omitted. The third demonstrated that wall paintings depicted monkeys in a variety of settings, two of which contained obvious religious elements paralleled in glyptic scenes. This section, then, examines the extent to which the monkey's roles in Egyptian religion and culture were adopted by the Minoans and the ways in which its roles were altered to suit the needs and pre-existing traditions of this Aegean culture.

As discussed in the previous chapter, baboons were associated with a number of divinities, particularly the sun god, Re, in their role as *bau* of the east and Thoth, who came to play an important role as advocate of the deceased. In non-religious roles, monkeys were associated with vessels and cosmetics. Tomb chapel scenes and New Kingdom ostraca often depicted monkeys performing a variety of human tasks. Scenes of music and dance were particularly common, but baboons were also shown picking fruit. Aspects of these functions can clearly be seen in Minoan art.

The first category of seals and the painting from Sector Alpha both reflect a remnant of the hamadryas' role as sun adorant or eastern *ba*. This was one of the earliest roles assigned to baboons in Egypt, and the MM II ivory seal from Mochlos attests to the Minoans' relatively early awareness of the association. However, while the baboons and other monkeys are depicted in the same attitude of adoration with up-raised arms, no trace of solar cult is present in any of the Minoan scenes. The object of the monkey's adoration was replaced with divine figures, as is the case of the monkey worshipping the seated goddess in the Kalyvia ring, or with a sacred structure, such as the altar or shrine in the fragment from Sector Alpha. Marinatos has also suggested that a second monkey, whose tail is visible in the existing fragment, was actually depicted inside the shrine (1987b: 417-419). If this is so, the Aegean artist may also have been aware that sacred baboons were kept within temple precincts at Thoth's main cult centres. Minoan style frescoes from the Hyksos capital of Tell el Dab'a do reflect an Aegean presence in Egypt either slightly before or contemporary with the creation of the Theran paintings. The artists may not, however, have understood the distinction between the hamadryas' associations with Re and Thoth. Additionally, they seem to have associated a variety of species with this gesture, so the Egyptian categorization of species was either misunderstood or ignored.

Next, the Xeste 3, room 3a painting depicts a monkey acting as an intermediary between the human and divine, a role commonly associated with Thoth in Egyptian religion. The significance of the ground floor painting on the north wall helps to explain this role. The girls involved are, like those on the first floor painting, all of different stages of development, as signified by their hairstyles. The girl to the right is in the second stage of development, while those to the far left and centre have long hair, signifying the fourth stage. The back lock which the girls had been growing from childhood is looped to the main strands with a pin, but that belonging to the central figure seems to have fallen out of place when she injured her foot (Davis 1986: 401). As noted in the previous section, the crocuses link this painting to the one on the first floor, where young girls are engaged in picking the flowers for saffron. Marinatos comments that, in addition to the product's use as a spice and dying agent, saffron is also a sedative used to relieve birth and menstrual pain, which leads to the suggestion that the blood trickling from the girl's foot is symbolic of blood lost during menstruation, childbirth, or most likely, the loss of virginity after a wedding (1987: 132). The last suggestion finds support in the hairstyle of the goddess, who retains her back lock of youth, indicating that she may be a virgin goddess. The girl whose foot is injured may be on the verge of marriage, which is signified by the release of her back lock during the trauma (Davis 1986: 406). The offering of saffron that the monkey makes to the goddess, then, is on behalf of the girls.

Marinatos' assertion that, because of this intermediary role, seal stones occasionally depict monkeys being worshipped by humans seems likely to be correct (1987: 127). Both the "Young Minotaur" and Zakro sealings show the seated animals facing standing human figures with up-raised arms. However, Marinatos' use of the term "semi-divine" to describe the monkey's status is questionable, since it is impossible to completely understand the monkey's position in Minoan religion. Thoth was one of the best loved and most powerful gods in Egyptian religion, and the fact that he helped humans through an important liminal period, the transition between life and death, did not make him less divine. However, it is possible that the Minoan monkey, who seems to aid young women through a different transitional state, may not have been assigned the same importance. The monkey does appear to have been considered more powerful than humans, but its relation to other divinities is unclear, despite portrayals of the monkey adoring a figure believed to be a goddess. Vignettes from the *Amduat* depict all the gods rejoicing with up-raised arms as the sun rises in the form of Khepri. Their attitudes of adoration and rejoicing do not make them semi-divine, just less important than the god to whom they make the gestures. Therefore, the gesture may allow us to make suggestions concerning the animal's importance in relation to other figures, but it does not allow us to judge whether the monkey is a divinity or a daimon. Interestingly, though, Aruz has remarked that the Xeste 3, room 3a scene may have inspired the creation of a similar scene on a hematite Syrian seal in Vienna. There young warriors, with the long hair, wasp-waists and broad shoulders and wearing the loincloths typical of Minoan youths, approach a seated goddess. Behind her a winged griffin stands on a short platform in much the same manner as the one in the Xeste 3 painting (Aruz 1995: 14-15). Although the monkey is absent, the similarities are striking. The depictions lend support to the Theran woman's status as a goddess. The Syrian artists certainly accepted her as such. The fact that monkeys appear in other Near Eastern seals with deities demonstrates that the monkey's association with divinities was common to Egypt, the Aegean and the Near East.

Of course, not all monkeys portrayed in either Minoan or Egyptian art have religious functions. It seems probable that the monkeys from the House of Frescoes, Xeste 3, room 4, and the Saffron Gatherer do not represent any form of divinity. Thoth had a baboon manifestation, but obviously not all baboons were considered divine. However, because baboons shared certain characteristics with divinities or engaged in behaviour that was perceived as religious, they were treated with respect and some were considered sacred. As discussed in the previous chapter, baboons were believed to be endowed with the secret language of the gods and only understood by the Pharaoh. They were also considered the most religious of animals because they appeared to rejoice at sunrise. Perhaps a similar role can be applied to the Saffron Gatherer, an apparently ordinary monkey, perhaps even a pet, that shares a divine association with saffron. The common Egyptian depictions of monkeys helping humans in their work may also have contributed to the image of the monkey picking stigmas from the crocus, which seems to have been an entirely Aegean invention.

Similar Egyptian images of monkeys playing musical instruments and dancing probably also contributed to the creation of musicians and sword fighters in the fresco fragments from Xeste 3, room 4, despite Marinatos' suggestion that, because room 4 is connected to room 3a by pier and door partition, the monkeys may be holding ritual contests in honour of the goddess (1984: 113). While this is possible, it would be a great departure from the Egyptian use of monkeys as entertainers for humans. Equally, no parallels exist for such a function in Near Eastern art.

In landscape scenes, monkeys appear at their most natural. They make no gestures of worship and only pursue activities suitable to monkeys. Those from Beta 6 are highly anthropomorphized, but there is nothing out of the ordinary about monkeys who enjoy swinging and jumping. The only truly unusual aspect to the painting is the fact that one monkey is shown *en face*. Morgan has argued that in Aegean glyptic scenes, this is symbolic of death (1995: 135-151). However, monkeys are never shown being sacrificed, and only the "Young Minotaur" sealing depicts a couchant animal, perhaps intended for sacrifice, in a scene with a monkey. Morgan does acknowledge that in Egypt, the frontal face is more commonly associated with birth and music than death (1995: 135). The god, Bes, for example, is usually depicted *en face*. This portrayal, then, would seem to follow the more common Egyptian use of frontal face symbolism. The abundance of marshland plants and eggs in the painting from the House of Frescoes would indicate a further association between monkeys and fertility. Even if the monkeys are eating the eggs, the idea of birds nesting would indicate a period of fertility. Additionally, in Egyptian art, marshlands are associated with the notion of fertility and rebirth. It should be noted, though, that the depiction of monkeys in landscape scenes appears to be an Aegean invention, since monkeys are rarely portrayed in any natural setting in Egyptian art without a human presence.

Finally, then, one may conclude that the Minoans did import many elements of Egyptian symbolism and meaning along with the image of the monkey. It was adopted as an intermediary between the human and divine realms and portrayed both as an adorant and as an object of adoration. It also seems to have been associated with women and fertility, not through cosmetic jars and the animal's maternal instincts as in the Egyptian tradition, but through one of the Aegean's most valuable commodities, saffron. They seem to have been enjoyed both for their natural behaviour and playfulness as depicted in the landscape scenes and their apparent ability to imitate and entertain humans. Interestingly, after LM IA, monkeys were no longer represented in Aegean art. Although Egyptian objects depicting monkeys have been found in Mycenaean contexts, the image was never adopted into their iconography. Monkeys would not be represented in the Aegean again until the Geometric period.

However, as mentioned at the beginning of the chapter, there was a continued awareness of monkeys in the Aegean, as evidenced by the Egyptian representations of the animal that were found in Mycenaean contexts and the earlier Minoan trinkets that were kept as heirlooms. As mentioned in Chapter 1, fragments of an Old Kingdom pithemorphic jar, depicting a mother vervet and her infant, were found at Mycenae, so monkeys were known on mainland Greece from an extremely early period. Cline has also cited the discovery of two Egyptian faience monkey figurines at Mycenae and Tiryns, dating from the 15th century BCE. The figures are both blue and depict guenons (vervets) (1991: 30-31). Their discovery is important for two reasons. First, it demonstrates that blue faience figurines were imported to the Aegean and adds support to the suggestion that the convention of blue was inspired by Egyptian prototypes. Secondly, it provides evidence that the Mycenaeans made a conscious decision not to adopt the image. The monkey's absence from Aegean art for the next several hundred years was not the result of a decrease in trade with Egypt and the Near East. The Mycenaeans were, apparently, not as impressed with the animal.[18] As demonstrated in the next chapter, when monkeys did appear in the Aegean again, their importance was much diminished. Eventually, the poor creatures would be despised.

[18] These are not the only Egyptian monkeys found in Greece. Cline (1991) provides a good overview of other examples found in Mycenaean contexts.

Catalogue of Aegean Seals, Sealings and Jewellery

Seal or Object	Provenance	Date	Form, material and measurement	Description of Scene or Object
CMS V.1, 233 **Fig. 63**	Chania Kastelli	LM IIIB	Impression Lentoid Dia. 1.2 cm	Two monkeys, leaning forward slightly, stand over a central object, possibly a plant. Their arms stretch out over the object, and their tails hang down. The monkeys are stylized, but their long muzzles indicate that they are baboons.
CMS II.1, 435 **Fig. 64**	Trapeza, cave	EM II- MM I	Pithemorphic stamp seal, Ivory. H. 4.4 cm Seal surface: Dia. 1.6 cm	A long-tailed monkey sits on top of the knob seal in a human pose. It is supported by its tail, and its hands rest on its knees. The surface of the seal is composed of an angular X-shaped pattern.
CMS II.1, 249	Platanos, Tholos tomb A	EM II- MMII	Pithemorphic Ivory seal. H. 3.5 cm	A stylized monkey, probably a baboon, sits with its arms between its legs. The surface depicts three stylised lions and an S-shaped pattern.
CMS II.1, 20	Ayia Triada, Tholos tomb A	EM II- MMII	Pithemorphic Ivory Seal. H. 1.8 cm	A stylized monkey, probably a baboon, sits with its arms between its legs. The base is roughly elliptical, with a cross-hatch design. The figure has lateral boring through the side for suspension, and the eyes are also drilled.
CMS V, S.1A, 302	Unknown	MM?	Pithemorphic seal, Ivory L. 1.0 cm W. 0.76 cm H. 1.77 cm	This is a seal carved in the shape of a seated monkey, probably a baboon. Its hands rest on its knees, and the tail seems to curve around the base. A hole is drilled through its side. The pattern on the base is composed of three parallel fish.
CMS V, 1A, 303	Unknown	MM?	Pithemorphic seal, Ivory, upper body missing.	Although the upper portion of this seal is missing, the lower part resembles the early Egyptian representations of seated baboons. The arms hang a little awkwardly from the shoulders so that the hands rest evenly between the feet.
CMS X, 30	Unknown	EM- MMI	Ivory/Bone Theriform ring Base: L. 3.7 cm W.1.4 cm Th. 0.4 cm Ring: H. 2.3 cm Outer Dia. 3.7 cm Inner Dia. 1.5 cm	This is an animal-shaped ring, perhaps a monkey, although a number of other identifications are also possible. The base is roughly rectangular, and the ring was carved around the natural hollow of the bone. The animal's head dominates the front of the ring, and two incisions form a v-shaped tail at the back. The seal pattern is composed of six sets of concentric circles.
CMS VIII, 109 **Fig. 66**	Unknown, Robert Erskine Collection	MM II	Pithemorphic Amethyst figurine. H. 2.3 cm	A stylized monkey sits in a crouched position. The right hand rests under the chin, while the left hand rests on the left knee. The tail is raised flat against its back. The face is rather flat and highly stylized, but the hair around the head suggests the mane of a baboon.
Karetsou 2000, no. 173 **Fig. 67**	Knossos	MM IB-II	Rock crystal pithemorphic figurine. H. 1.7 cm	A monkey with a highly stylized face sits in a crouched position. Its arms and hands are not visible. Incisions along the chest and stomach may indicate a shroud.
Karetsou 2000, no. 174 **Fig. 68**	Koumasa, Tomb B	MM	Lapis lazuli pithemorphic figurines/amulets	Two crouching monkeys, highly stylized, with attachment holes drilled through their heads. The pair seem to have been attached to necklace with lapis lazuli beads.
Athens National rchaeological Museum, no. BE 1996/11.12 **Fig. 69**	Unknown, Aidonia Treasure	LM IA	Electrum finger ring with pithemorphic bezel. Dia. of ring 1.6 cm Dia. of bezel 1.3 cm	Repoussé bezel of fine gold sheet in the form of a squatting, tailless monkey. The simian, in profile, raises its hand to its mouth as if eating.
Higgins 1979, nos. 17, 64, 65 **Fig. 70**	Unknown, Aegina Treasure	LM	Four gold ornaments with cornelian beads. Dia. 6.5 cm	Two sets of ornaments, probably earrings. Within a ring in the form of a two-headed snake, a pair of greyhounds stand facing each other, with their muzzles joining. Each lifts a forepaw, which touches that of the other dog. The wear wire collars attached to leads, which have been threaded through cornelian beads and attached to the inner side of the ring. The dogs are supported by a pair of crouching, stylized monkeys that hold their hands to their mouths, as if eating. Both monkeys and dogs appear to be wearing belts. Attached to the outside of the ring, are fourteen chains with alternating disc and owl ornaments. The chains with owls are strung through elongated cornelian beads attached near the ring. A small cornelian bead is attached at the end of every chain above the owl or disc. The entire piece is hollow and reversible.

Seal or Object	Provenance	Date	Form, material and measurement	Description of Scene or Object
CMS II.1, 473 **Fig. 71**	Mochlos, grave	EM II-MM I	Ivory Pear-shaped seal with a perforated eyelet. Seal surface: Dia. 1.1 cm	Two stylized baboons sit back to back. Their arms are raised at the elbows, but the tails are lowered. An incised line is apparent above their heads.
Karetsou 2000, no. 162	Knossos	LM IA	Ovoid seal impression. H. 1.6 cm	A stylized monkey seated on a campstool. Its arms are raised at the elbows, and its tail is raised at the back. It wears a belt around its waist, and faces a human figure, partially in lacuna.
Giamalakis 372	Prassa	LM	Chalcedony Amygdaloid H. 3.3 cm W. 1.7 cm	A male figure, wearing a loincloth, and a baboon face each other. The baboon, left, is seated in profile with its right forearm (or both) raised. The male figure, centre, also raises both arms at the elbow. His face is shown in profile, but his torso is *en face*. A vertical series of plants line the left side of the impression. In the background, wavy lines are shown to the right and left of the male figure. The scene is framed above and below by double horizontal lines.
CMS II.7, 24 **Fig. 72**	Kato Zakros House A, Room 7	LM	Impression Lentoid Dia. 1.4 cm	A female figure, wearing a flounced skirt, stands on the far left-hand side of the scene, facing a large, seated baboon to her right. Her left arm is raised at the elbow, while her right rests on her chest. The baboon faces her with both arms raised at the elbow. Its tail is raised behind its back.
Giamalakis 359	Siteia	LM	Steatite Lenticular Dia. 1.4 cm Very worn, in two pieces	A seated monkey, perhaps a baboon, left, faces a female figure, who occupies most of the scene to the right. The monkey holds a tall object, perhaps a pot or basket, and the female figure leans towards it slightly. Her chest is shown *en face*, and her arms are rendered in a stylised manner, forming an "m" from her shoulders.
CMS II.3, 103	Kalyvia, Tombe dei Nobili, Tomb 2	LM	Ring, gold with bronze core. Ring with a concave elliptical plate. Ribbed outer ring. Plate: L.1.7 cm W. 1.0 cm Ring: Inside Dia. 1.3 cm	The scene is divided into two parts. On the far left, a female figure clothed in a flounced skirt faces right, with her back to the viewer. Her arms are visible on either side of her torso, and both appear to be raised at the elbow. A baboon stands directly in front of her, also facing right, with its forearms raised. Their gaze seems to be directed towards the nude female figure to the right, who also has raised forearms. She, however, seems to be in a seated position, and her torso and face are shown *en face*. Behind her to the far right is a pillar. A number of motifs are depicted above and between the figures.
Evans PM II, Fig. 492	Ayia Triada	MMIII-LM I	Impression	A long-tailed monkey sits with its arms raised at the elbows. Two stylized plants float in front of the monkey, while a third hangs floats above its curved tail.
Hood 1978, Fig. 224b **Fig. 73**	Ayia Triada	LM I	Impression	Two baboons flank an altar. Their arms are held down in front of them, and they lean forward slightly. Their tails are raised behind them.
Giamalakis 355	Phaistos	LM	Chalcedony Lenticular Dia. 1.7 cm	Two monkeys, shown standing and in profile, flank a kantharos. The monkeys' tails are long and curled upwards at the ends. They stretch their arms horizontally over the top of the kantharos; their hands seem to meet over the centre. Both monkeys look backwards, their heads shown in profile.
Giamalakis 108	Mallia	MM	Chalcedony Quadrangular prism H. 1.1 cm W. 0.6 cm	Four separate scenes are depicted on this seal. 1) a hieroglyphic sign 2) a couchant bull 3)a seated baboon with upraised forearms, facing right 4) a swirled pattern
CMS X, 31	Unknown	EM-MMI	Ivory/Bone Pithemorphic seal L. 1.2 cm W. 0.9 cm H. 2.75 cm	This is a roughly carved seal in the shape of a crouched monkey. Its species is unidentifiable. The base is roughly elliptical, with a cross-hatch design. Lateral boring is visible below the monkey's head.
CMS X, 50	Unknown	MM	Agate Rectangular block, banded pink and yellow, burnt opaque L.1.25 cm W. 1.13 cm Th. 0.40 cm SH. 0.25 cm	Two baboons stand facing each other. Their arms project from their chests in a downward L-shape. The reverse is a pattern of six dots divided by three wavy lines.

Seal or Object	Provenance	Date	Form, material and measurement	Description of Scene or Object
CMS V, S.1A, 159	Chania	LM	Impression, Lentoid Dia. 1.15 cm	A seated baboon, shown in profile facing left, raises both arms at the elbow. One arm is slightly higher than the other. The baboon's tail is also raised. A Linear-A symbol hovers above the tail.
CMS V, S.1A, 131	Chania	LMI	Impression, Oval L. 1.1 cm W. 0.9 cm	Two highly stylized baboons sit facing each other. Both have their arms raised at the elbow with the palms apparently turned toward their faces. Their tails are curled under their feet.
CMS I Suppl., 114	Unknown	LM	Gold ring, Elliptical Plate: L. 1.84 cm W. 1.26 cm Ring: Dia. 1.7 cm Very worn	On the far left of the scene, an animal, perhaps a monkey faces a seated female figure in the centre of the scene. She wears a flounced skirt and appears to extend an arm towards the monkey. Behind her, to the far right, is a palm tree.
CMS II.7, 114	Kato Zakros	LM	Impression Pyramid Fragmentary	A seated baboon, in profile, raises at least one arm, which rests on its knees.
CMS I, 377	Pylos palace	LM	Impression Lower left section missing	A male figure, left, faces a baboon to the right. The man's face is in profile, and only his upper torso remains in the fragment. The baboon stands with its face in profile and its back is shown in ¾ profile. Both arms are raised at the elbow and visible on either side of its body. The baboon's head is level with the man's shoulders, and it wears a belt. Its tail is raised behind its back, and it appears to make a step towards the man with its left leg.
CMS I, 478	Crete?	LM	Carnelian Lentoid Dia. 1.25 cm	A highly stylized seated baboon, shown in profile, raises both arms at the elbow. It faces toward linear symbols to the right. Its tail curls behind it in two alternating crescents.
CMS II.5, 297	Phaestos, room 25	LM	Impression Round, slightly concave Dia. 1.1 cm	A stylized, seated baboon raises its arm at the elbow. The hand reaches outward slightly with the thumb separated from the palm. The baboon's tail is raised behind its back. A branch is can be seen between the monkey's hand and foot and another between its tail and back.
Karetsou 2000, no. 170	Platanos, Tholos tomb A	EM III-MM IB/ IIA	Hippo ivory, double pithemorphic figurine. H. 3.7 cm W. 4.2 cm	Two stylized monkeys sit back to back. Their limbs are indicated in low relief on their slightly triangular bodies. Their eyes are drilled holes.
Karetsou 2000, no. 172	Ayia Triada, Tholos tomb A	EM III-MM II	Ivory pithemorphic figurine. H. 2.3 cm	This highly stylized seated monkey can readily be identified as a baboon because of its elongated muzzle. Its feet are well articulated, but its arms are crudely rendered. It has the appearance of being unfinished.

Catalogue of Aegean Wall Paintings

Provenance	Date	Description
Knossos, The Early Keep **Fig. 76**	MM IIIB-LM IA	The Saffron Gatherer—This is a highly fragmentary painting. The main fragment reveals the leg, arm and midsection of a monkey's body. It is blue and wears a red arm band, a bracelet, a halter on its chest and a belt around its waist. Its slender hand reaches out to touch or pick a crocus growing from a pot. A fragment of tail has also been recovered, along with a flesh coloured muzzle that must belong to a second monkey. The extant fragments reveal that the monkeys are depicted as slender and long-tailed. They are not anthropomorphized.
Knossos, The House of Frescoes	LM IA	The Monkey and Bluebird Frieze—This fragmentary fresco has been restored to depict six monkeys in a naturalistic, but not realistic, landscape with blue doves flying overhead. The landscape displays a mixture of plants, including ivy, papyrus, lilies and crocuses, and a watercourse runs through the scene. Most of the monkeys have been reconstructed from small fragments, but one well preserved simian bends its arm awkwardly around a papyrus stalk. Another was reconstructed eating an egg from a nest. The monkeys have slender bodies and show no signs of anthropomorphization.
Thera, Beta 6 **Fig. 74**	LM IA	Eight large monkeys jump and swing through a rocky landscape. The simians have vervet-like faces, long curling tails and human bodies. One of the monkeys is shown *en face*, but the rest were painted in profile. No vegetation is present in the surviving fragments, but a blue area below the monkey would indicate a watercourse.
Thera, Xeste 3, Room 3a **Fig. 75**	LM IA	The monkey in this scene faces a woman seated on a tripartite structure. It stands with its right foot on the ground and its left on the first level of the platform. It holds a clump of saffron in its hands, which it seems to be offering to the seated woman. A winged griffin stands on a section of the platform behind her. The woman has saffron on her cheek and dress, which has a blue border decorated with crocus flowers. She also wears necklaces adorned with ducks and dragonflies. Her hair is pulled back with a blue band, and a curled lock at the back, decorated with beads, rises above the back of her head. Behind the monkey, a girl empties her hand basket of crocus flowers into a larger basket on the ground. She looks up at the scene before her. The rest of the scene, which covers two floors, depicts girls of various ages picking crocus flowers. One of the older girls sits on a rock, holding her wounded foot. Her hairpin has fallen off, releasing her back lock.
Thera, Xeste 3, Room 4	LM IA	This scene is highly fragmentary, but one monkey can clearly be seen playing a stringed musical instrument, and another brandishes a sword that it seems to have pulled from a scabbard. Another monkey faces the tip of the sword. Most of its face and both its hands are in lacuna, but its arms seem to be outstretched, as if holding something, possibly another sword. At least one of its knees is drawn up toward its chest, touching its elbows. The monkeys have patas-like faces, with almond eyes. Their bodies are anthropomorphic.'
Thera, Sector Alpha	LM IA	This is a highly fragmentary scene depicting a seated blue monkey with arms upraised at the elbows, facing an architectural structure, probably a shrine, supported by a column with a papyriform capital and topped with horns of consecration. A fragment of the monkey's long curling tail also survives, and the tail of another monkey can be seen on the other side of the column. The monkey's face is vervet-like, but its body has human curves. Its hands are poorly formed. The right hand resembles a club, and the left is barely visible.
Phylakopi, Pillar Crypt	LMIA	This small fragment depicts a monkey's eye and the top of its blue head in profile. Nothing is known of the scene from which it came.

Chapter 4
MONKEYS IN THE GRECO-ROMAN WORLD

After the period of Minoan cultural pre-eminence in the Aegean, monkeys disappeared from Crete, Mainland Greece and the Cycladic Islands. Only Egyptian examples and earlier Minoan trinkets were found in Mycenaean contexts. By the Geometric Period, however, monkeys began to make their comeback in the form of bronze figurines, which may have been inspired by Egyptian amulets. Later, monkeys appeared in fables, vase paintings, terracotta statuettes and perfume containers, and even wall paintings. Unfortunately, though, the Greeks did not view simians with the same respect they were accorded in Egypt and Minoan Crete. The Greeks did not create images of their deities with animal attributes, and unfairly ridiculed the Egyptians as 'animal worshippers.'[19] The perceived worship of monkeys, in particular, was held in contempt because the animals themselves were considered foolish, vain and, perhaps because of the anthropocentric nature of this culture, ugly.

The Romans were just as harsh with our primate cousins. On a few occasions, monkeys were even included in the arena. Because of their obvious similarity to humans, though, a number of philosophers, doctors and natural historians from both cultures became interested in their anatomy as a means of understanding the human body. Aristotle was probably the first to examine the animal's physiognomy, and Galen later vivisected them (*AA* 9.18-19). A few monkeys, however, were lucky enough to be kept as beloved pets. In Greco-Roman Egypt, baboons continued to be associated with Thoth, but the increased popularity of animal cults during these times meant that hundreds of monkeys and hundreds of thousands of ibises were sacrificed to him. Barbary macaques seem to have been imported from the western shores of North Africa to substitute for the hamadryas, whose population must have been dwindling as a result.

Because of the vastness of the Greco-Roman world, the amount of cultural transference and the immense time span, the cultural values of monkeys cannot always be assigned to particular time periods or regions. Many artistic and literary categories overlap, and a highly ambivalent attitude toward the animal is apparent in a number of works. The creature's image inspired reactions of curiosity, mistrust, delight and disgust. Its form was strange, but its resemblance to humans was (and is) unmistakable. Sadly, its proportions did not match the Greco-Roman ideal, and its playful foolishness led to a mixture of warm-hearted laughter and cruel scorn. The monkey was generally considered an inferior imitation and imitator of man, but occasionally, it displayed a curious glimmer of wisdom.

Unlike the other chapters, therefore, this one is not divided into sections. Instead, it examines the reception of real monkeys and the handling of their 'cultural baggage' as both the animal and the image were introduced to various areas of the Greco-Roman world from the Near East, Egypt and western North Africa. Certain associations and practices were limited to particular regions and time periods, but most artistic representations and literary references to monkeys reflect recurring themes in the Greco-Roman attitude toward monkeys. Of course, this chapter cannot cover every encounter the monkeys had in their 1,000 year relationship with the Greeks and Romans, but hopefully, it will capture the highlights and correct a few errors of modern scholarship along the way.[20]

As mentioned above, monkeys began to emerge in Greek art during the Late Geometric period in the form of recognizably simian figurines as well as stylized free-standing anthropomorphic bronze figurines. Langdon (1990) catalogued twelve known examples of the latter and convincingly argued their origin and possible significance, and her interpretations are worthy of discussion here. All

[19] Cicero, *Tusc.* 5.27. 78, ridicules animal worship without mentioning monkeys (King, Trans. 1927):
Aegyptiorum morem quis ignorat? quorum imbutae mentes pravis erroribus quamvis carnificinam prius subierint quam ibim aut aspidem aut felem aut canem aut crocodilum violent, quorum etiam si imprudentes quid iam fecerint, poenam nullam recusent.
Who does not know the custom of the Egyptians? Their minds are infected with degraded superstitions and they would sooner submit to any torment than injure an ibis or asp or cat or dog or crocodile, and even if they have unwittingly done anything of the kind there is no penalty from which they would recoil.
Juvenal, *Satires* 15.4, adds long-tailed monkeys (not baboons!) to the list: effigies sacri nitet aurea cercopitheci.

[20] This chapter follows the development of simian iconography and literature from the late 8th century BCE, when the first representations of monkeys appeared at Greek sites, up to the early Christian period.

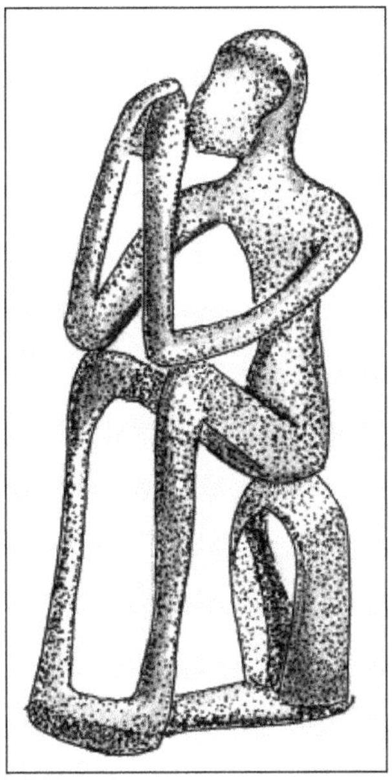

Fig. 77: Bronze figurine, possibly from the valley of Alpheios (Baltimore, Walters Art Gallery)

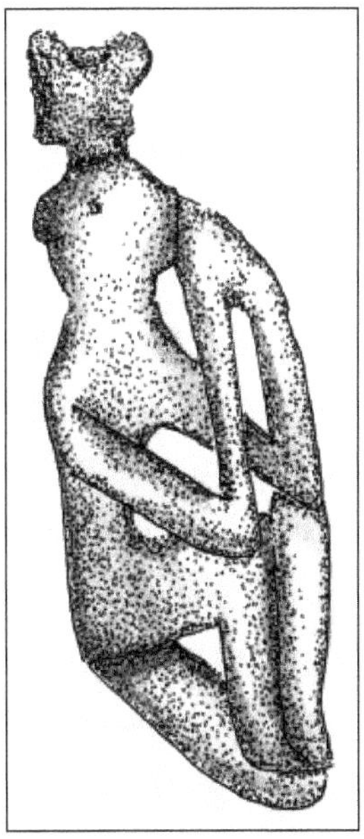

Fig. 78: Bronze monkey from the sanctuary of Athena, Kameiros (Archaeological Museum of Rhodes)

are seated figures, which either have their hands resting on their knees or lifted to their mouths. Figurines of the latter type sometimes hold objects up to their lips. Most are highly anthropomorphic and probably depict men, but some have simian features, such as elongated chins, heavy brow-ridges and pointed ears. Two examples are ithyphallic. Usually the figures are seated on stools, but a few are crouching on the ground. They have been found in such diverse areas as Elis, Laconia, Arcadia, Euboia, Rhodes and Central Italy. The few that have been found in recorded contexts come from sanctuaries: Olympia, Athena Alea at Tegea, Artemis at Mavriki, Artemis Orthia at Sparta, and Athena at Kameiros. Although even these cannot be securely dated, either because they were found in votive pits with objects of earlier and later periods or because the records were imprecise, Langdon maintains that their base forms are similar to those of thousands of bronze animal figurines, dating from 750-700 BCE, found at Olympia (1990: 411-12). Their provenance suggests that they were intended for votive offering.

Schweitzer has argued that figures holding objects to their mouths (**Fig: 77**), are drinking from flasks and that they may be considered early satyrs or companions of Dionysus. He compares the figures to those seated on top of cylindrical, studded objects commonly known as bottle stoppers, which have been found in areas such as Macedonia, Chalcidice and Thessaly (1971: 160-1). These objects are slightly later in date. Bouzek dates a number of them to the middle of the seventh century (1974: 76-7), and Langdon notes that the earliest example comes from Kuci Zi, Albania, and dates from c.700 BCE. She also observes that the intended use of the objects has been incorrectly interpreted. The so-called jug or bottle stoppers have been found in situ at the waists of skeletons, suggesting that they were actually worn as belt pendants (1990: 414). The figures seated on the tops of these pendants also hold objects to their mouths, but, rather than flasks, the objects may be fruit or rudimentary musical instruments, such as conch shells. Related simian figurines, such as the bronze figurine from Kameiros that unequivocally depicts a monkey playing a flute, lend support to this interpretation (Langdon 1990: 413, fig. 11; **Fig. 78**). A number of seated figures on bird-cage pendants may also have been inspired by the seated figurines. Five of these pendants have been found with seated figures, but none comes from a recorded context. However, about two-thirds of the 61 known examples come either from graves or sanctuaries in northern Greece (1990: 414). Bouzek has remarked that some of the most realistic figures were broken off from the stopper pendants and saved (1974: 79-80). Langdon suggests that these may have had an amuletic function (1990: 415).

Although most scholars accept that the seated figures are men, Langdon argues that very few examples are entirely human in appearance and notes that some scholars have suggested alternative identities, including demons, satyrs, apes and grasshoppers.[21] She argues, however, that the figures should be viewed as human with simian components (1990: 415). She comments on the significance of the monkey through its first appearance in the Aegean during Minoan times, when it was adopted from images of Thoth, portrayed as a squatting baboon and adds that monkeys also had a long history in the Near East. During the period of Mycenaean influence monkeys disappeared from the Aegean, but made their comeback all over the Mediterranean between the 9th and 7th centuries. Langdon attributes the image's renewed popularity to the Phoenicians, asserting that the distribution of monkey figurines corresponds exactly with their trade routes at that time (1990: 419).[22]

The Phoenicians, in turn, adopted the iconography and significance of monkeys from a number of sources. The

[21] For the sources of the various interpretation, see Langdon (1990: 415, notes 34-38).

[22] Wisely, I believe, Langdon views the two periods as "discrete but related Orientalizing phenomena" (1990: 423), and does not suggest continuity of the image in the Aegean.

Egyptian dwarf god, Bes, was adopted by the Phoenicians, and as argued in the first chapter, he was closely associated with monkeys by this time. Additionally, she states that monkeys were imported to the Levant from India, and Syrian seals depicting seated monkeys with a nude goddess attest to the animal's association with sexuality and fertility in the Near East. Most monkeys at this time, though, would have been imported from Egypt. Assurbanipal II, for example, lists monkeys as part of the booty he took from Thebes (Dunham 1985: 235). Langdon remarks that more realistic monkey figurines found in Etruria were probably created directly from Near Eastern prototypes rather than from Greek imitations of Phoenician or Syrian amulets (1990: 419). Clearly, the inspiration was from a more direct source, but as will be shown later, the Etruscan monkeys were more likely inspired by North African macaques and the uniquely Punic interpretation of Syrian amulets.

Based on stylistic similarities, she suggests that the anthropomorphic figurines seated on bases, which were found predominantly in Greek sites, are of Laconian manufacture. The Northern Greek bottle-stopper and bird-cage pendants were, in turn, inspired by these seated figurines (Langdon 1990: 422). Langdon notes that the ithyphallic nature of some of the bronzes was not portrayed in the Aegean monkeys of the Bronze Age and suggests that this development may have been inspired by foreign, particularly Egyptian, models. This connection to fertility made them acceptable votive offerings in sanctuaries, and it seems possible that the Tree of Life may be represented by the stylized rod of the bottle-stopper pendants.

Langdon reasons that Greeks of this period would have been familiar with monkeys mainly through travellers' tales and artistic representations and that they may have been impressed with the lore attached to these animals, particularly the human qualities that would have been attributed to them. The mixture of simian and human characteristics was, then, a conscious decision, designed to imbue the figures with the desired qualities associated with monkeys. She maintains that this reverence of monkeys ended in Greece by the beginning of the 7th century, when the creatures were employed in a satirical fashion in a variety of other media and that the last remnants of the period survived in the pendants produced in Northern Greece (1990: 423).

For the most part, I agree with Langdon's suggestions, but I suspect, and will attempt to demonstrate further on in the chapter, that a certain reverence for monkeys lingered in a few areas of Greece longer than she realises. Certainly, though, Langdon's argument that the Phoenicians were largely responsible for the monkey's renewed popularity around the Mediterranean is entirely correct, and their influence in the production of simian imagery can be seen in variety of media. It must be noted, however, that objects manufactured on the Phoenician mainland are often difficult to distinguish from imitations created at sites abroad, particularly in areas such as Cyprus, Rhodes, North Africa, and Sardinia.

Furthermore, the Phoenician culture itself comprised elements borrowed from the numerous peoples with whom they traded. The issue is further complicated by the possibility of Phoenician craftsmen working in Cyprus and other areas, who may have emphasized particular motifs to correspond with local political ties. Markoe notes that Phoenician-style bowls of Cypriot provenance demonstrate strong Assyrian influences and that this is in keeping with the fact that Cyprus was a tributary of King Sargon II (1985: 8).

Two of these Cypro-Phoenician bowls, dating from c. 710-675 BCE, represent an important piece of the Mediterranean monkey puzzle. One silver bowl was found at the Etruscan site of Praeneste and the other at Kourion in Cyprus. The bowls are similar but not identical, and each is decorated with an outer zone depicting the same narrative of an "ape" hunt, which takes place in nine consecutive scenes: a prince, king or hero leaves his city in a chariot; he stops to shoot a stag; he pursues the wounded animal; he flays the stag, which hangs upside down from a tree, while his charioteer looks after the horses; the prince makes an offering to a winged deity, while an ape steals something from the offering pile; the ape attacks the prince, but the winged deity rescues him by lifting his chariot away from the scene; the prince is back on the ground and able to pursue the ape; he kills the it; and returns to the city in triumph (**Fig. 79**).

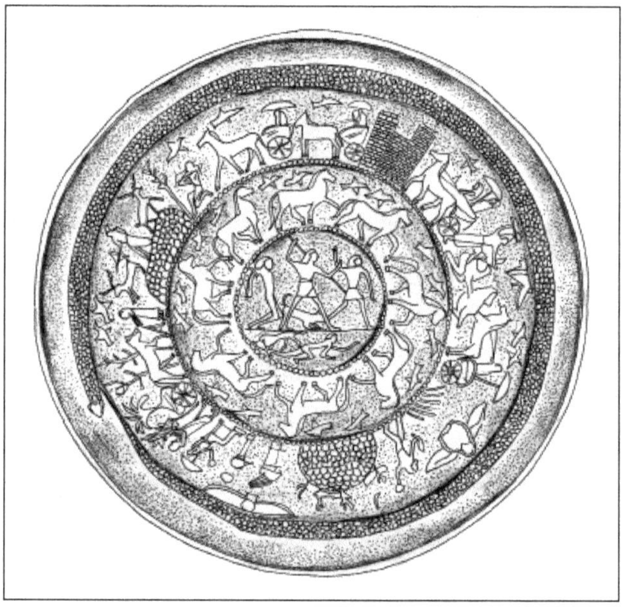

Fig. 79: Silver bowl from Praeneste, depicting an 'ape hunt'

The bowls have aroused much discussion concerning both the species of monkey portrayed and the significance of the narrative. McDermott calls the simian creature a gorilla (McDermott 1938: 234-5), in accordance with Clermont-Ganneau (1880) who was the first to make an in depth study of the Praeneste bowl. Markoe (1985) simply refers to the animal as an ape, without naming any particular species. Hermary calls it a monster or pseudo-gorilla (1992: 133, 136), and Marquand also described it as a monster that may have been modelled on a gorilla (1887: 332). The suggestion that the creature is a non-human primate-monster is most likely correct, as long as the primate in question is a monkey rather than a gorilla. The Phoenicians could not have come into contact with gorillas, since their knowledge of African primates came mainly from Egypt, Carthage and Tingitana. These cultures made no representations of great apes, and no faunal remains of great apes have ever been recovered from these sites. Hybrid gods and monsters composed of various animal parts, on the other hand, were not unknown in the Near East. A neo-Assyrian composition describing statues of Babylonian deities and demons even mentions one that has the face of a monkey, the ears of a sheep and human hands. Her body is combination of fish and dog (Dunham 1985: 258-9). Furthermore, monkeys would probably not have been a common sight in Cyprus, where the bowl was probably manufactured. Monkeys were never native to the island, and, as was the case with Sparta, the local people would mainly have known about them from travellers' tales. It is easy to imagine someone describing a large, thieving baboon or macaque to Cypro-Phoenician artisans, who may have portrayed these exotic creatures as monstrous giants.

The interpretation of the scene itself is more complicated. Marquand described the hero's garments as Assyrian but interpreted the charioteer as Egyptian (1887: 329). He interpreted the winged divinity as a solar deity on the basis of Egyptian and Near Eastern representations of a winged solar disc and identified the hero as the legendary Cypriot king, Kinyras (1887: 331-335). Hermary (1992), on the other hand, views the hero as a form of Heracles derived from the Egyptian god, Bes, and the monster as one of his adversaries, such as Geryon, whose legend was well attested on Cyprus. This suggestion is based on a comparison with other Cypro-Phoenician bowls, which depict a hero in a lion-skin fighting an animal, in one case a lion, with his bare hands. As later literary evidence that Heracles was derived from Bes, Hermary cites a passage from Diodorus, 5.64.7, which describes the Cretan Heracles as a god invoked by women who also wear amulets made in his name. He is not troubled by the Assyrian dress of the hero in the "ape hunt" scenes because of the melange of styles found in Phoenician art and insists that the scene contains many elements from the myth of Heracles. The hero is attacked by a stone-throwing monster and rescued by a deity, just as Heracles was aided by Athena on his journey to the Centaur Pholos (Hermary 1992: 133-5). Markoe's suggestion is less detailed; he simply views the narrative as a lost Phoenician fable or epic (1985: 68). In fact, the two suggestions are not incompatible or improbable. It is around this time that Near Eastern/Egyptian fables, better known as Aesopic fables, began to appear in Greek literature. It is even possible that the ambivalent attitude toward monkeys that would emerge in later Greek culture was inspired by Phoenician tales of simian monsters. Before examining other negative depictions of monkeys, however, a survey of positive representations is in order.

During the 7th and 6th centuries, plastic vases, many of which were pithemorphic, began to be produced in Eastern Greek centres, particularly Rhodes, as well as Corinth and Etruria, and most of those produced after the second half of the 7th century were intended for use as scent-bottles. The Rhodian examples were usually mould-made, a technique reintroduced from the East in the late 8th century, while Corinthian and Etrurian vases were usually wheel-made

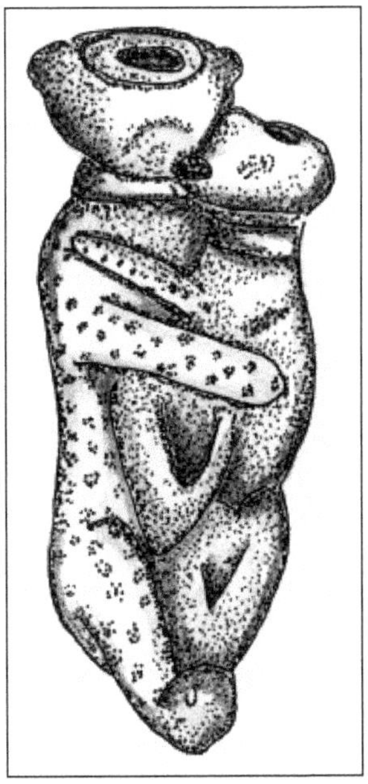

Fig. 80: An Etrusco-Corinthian aryballos (After Walker 1996: pl. 15, no. 90)

vessels with hand-modelled details (Higgins 1959: 7). During this period, it seems that the production of scent-bottles became such a great business on Rhodes that the terracotta industry was eclipsed and temporarily ceased to exist, and Etruria seems to have been the main consumer. Aryballoi found on the Greek mainland are more likely of Corinthian manufacture (Higgins 1959: 9-10).

The Leo Mildenberg collection contains several such pithemorphic aryballoi. One Etrusco-Corinthian example, dating from the early 6th century, is in the form of a mother and her infant (**Fig. 80**). The heads and bodies of both figures are typically hollow and wheel-made, while the limbs are solid and hand-modelled, as are the ears, muzzles and ischial callosities. Both animals wear collars around their necks, and both are resting in a squatting position, with the knees drawn up and one foot crossed over the other. The mother clasps the infant tightly to her chest with both arms. The infant's bottom rests just above the mother's left knees, and its feet touch her right knee. Like most of the Etrusco-Corinthian pithemorphic aryballoi, the features are highly stylized. The heads are flattened and perforated with a circular hole. The hint of fur is created with rows of black dots along the bodies and limbs of the two monkeys, while the heads, collars and hands are black (In Walker 1996: 64, fig. 90). Another aryballos from the same collection is actually mould-made with hand-modelled details. Like the previous example, this monkey sits on its callosities with its knees drawn toward the chest and its feet crossed. The artist, however, demonstrates a sense of humour: the monkey holds its left hand over the top of its muzzle as if in disbelief (**Fig. 81**; In Kozloff 1981: 114, fig. 95). The gesture is similar to that of a 7th-6th century BCE Eastern Greek or Cypriot baboon figurine, who sits with his right hand over his muzzle. His deeply incised lips suggest a mournful, exasperated expression (**Fig. 82**; In Walker 1996: 62, fig. 89). Both the mother/infant and comical aryballoi were probably designed to be worn, since in both cases, the feet extend too far forward

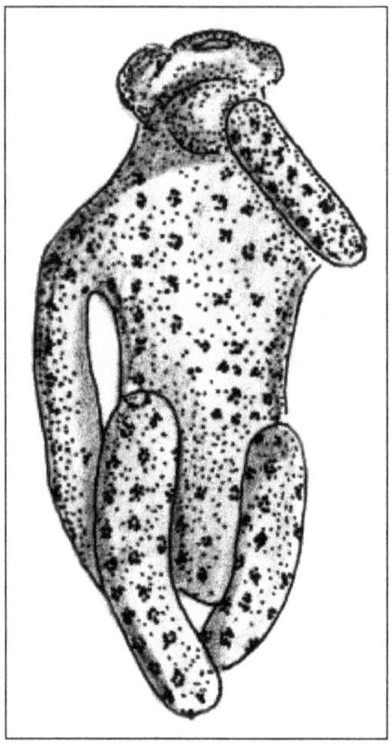

Fig. 81: An aryballos
(After Kozloff 1981: 114, no. 95)

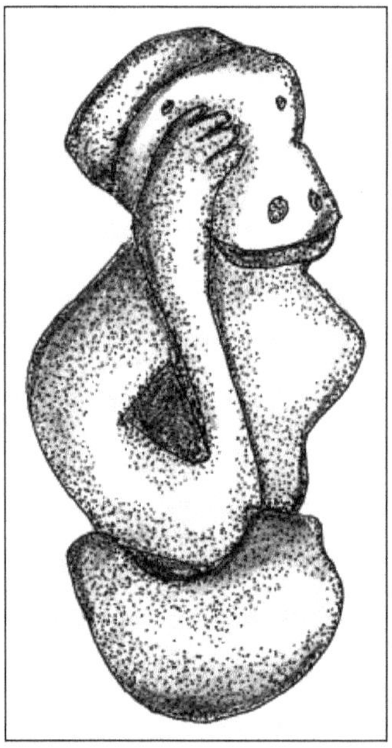

Fig. 82: Eastern Greek or Cypriot baboon
(After Walker 1996: cover, no. 89)

for the vessel to stand upright. Some 6th century BCE Rhodian aryballoi, on the other hand, are designed to stand upright. A small number of vessels depict monkeys in a highly abridged form, without arms or legs. These seem to be related to other truncated-animal aryballoi, such as antelope vessels, and all were probably produced in the same workshop (In Kozloff 1981: 117, fig. 98; **Fig. 83**). The monkeys' heads are more realistically shaped and decorated with painted and incised features. The elongated, round muzzle is defined by an incised mouth, and the curved stripes around the edges of the monkey's lips create the impression of a smile. The almond-shaped eyes and old-world monkey nostrils were painted with considerable care, and the rounded ears are painted black.

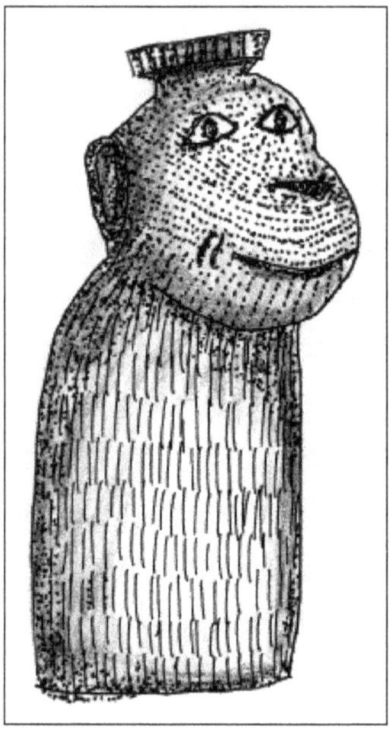

Fig. 83: Rhodian aryballos
(After Kozloff 1981: back cover, no 98)

The head is further decorated with rows of black dots, probably to depict short hair, while the fur of the tubular body is painted with bands of short lines. The bottom of the vessels is flat and decorated with a rosette pattern. Towards the back of the monkey's neck, the vessel curves forward to suggest hunched shoulders.[23]

The ultimate inspiration for these pithemorphic aryballoi is most likely Egypt, where perfume and cosmetic containers were produced in the form of monkeys or with monkeys in relief from the Old Kingdom onwards. A stylized 8th-7th century BCE bronze monkey, once again from the Leo Mildenberg collection, may represent an intermediate step in the transition from the production of bronze figurines to terracotta aryballoi in Etruria and mainland Greek centres. This monkey sits in a crouched position with its feet spread apart. The head faces backwards at an angle of 180° from the body, while the arms form a circle in front of the body. The little creature was probably originally designed to hold a vessel, such as a kohl pot. The object does not seem to have come from a recorded context, but Mitten remarks that the features are more obviously simian than the Peloponnesian and Northern Greek seated figurines, discussed above (In Walker: 61, fig. 88). It seems likely that the idea of monkeys as exotic animals associated with perfume, women, and perhaps even protection was imported from Egypt via Phoenician traders at approximately the same time the bronze simian-like figurines were incorporated into Greek culture as votive offerings associated with fertility. The probability that some of the Etrusco-Corinthian aryballoi were intended to be worn suggests to me that they may have had some amuletic function in addition to their primary use as vessels. This seems particularly probable in the case of the mother/baby aryballos, which was most likely created for a female wearer. I would suggest that popularity of terracotta

[23] In addition to the Leo Mildenberg example, a similar truncated monkey aryballos is located in the Museum of Fine Arts Boston, Reg. no. 192.

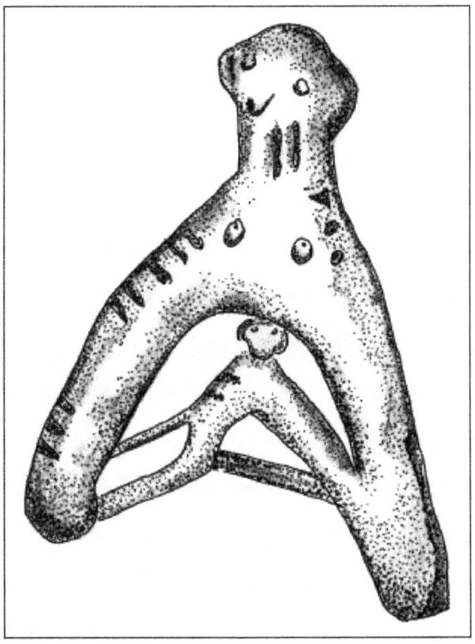

Fig. 84: A Boeotian monkey and baby figurine from the Leo Mildenberg Collection (After Kozloff 1981: cover, no. 92)

containers in Eastern Greek and Etruscan sites, slightly after the main period of bronze figurine production, may have led to a certain democratization of the simian image and significance. The perfumed oil contained in the vessels could have been an expensive, exotic substance, but the vessel itself was probably not costly to produce. Although these aryballoi do not truly fit the common use of the term terracotta, employed to signify figurines, their moulded elements demonstrate many similarities with toys and votive offerings.

A type of terracotta that may have been employed for both purposes is the stylized Boeotian monkey figurine. These figurines depict monkeys in a seated position, with their short legs out straight. The palms of their hands rest on top of their feet or, for a more accurate description, the ends of their stump-like legs, so that their bodies have a triangular appearance (**Fig. 84**). The heads are also roughly triangular, and both the head and body are decorated with crude, uneven horizontal stripes. A few figures of this type have a similarly decorated infant seated within the body of the mother, while others, with or without an infant, have small lumps attached to the chest to indicate breasts. The rest are devoid of sexual characteristics (Preston 1975: 125). In a brief study of type, Preston found that of the approximately twenty-six known examples only thirteen came from recorded contexts, either at Rhitsona or Tanagra. Those from Tanagra cannot be securely dated, but a number of figurines from Rhitsona were found in early to mid sixth century BCE graves. As Preston maintains, the evidence contradicts the assertions of some earlier scholars, who argued for an earlier date in the Mycenaean or Geometric period (1975:121). Robinson even described a mother and infant figurine, reportedly found near Mycenae, as a 'bitch with young puppy' and dated it to LH III (1934: 28). Mollard-Beques dated an unusual example of a mother and baby figurine, in which the infant rides on the mother's shoulder, to the 8th century BCE (1954: 26, no.B155). She did, however, recognize the animals as baboons. Curiously, she dated a similar, although noticeably cruder, figurine without an infant to the early 6th century (1954: 4, no. B10). Preston maintains that even the four figurines in the J. Paul Getty Museum differ considerably in quality, remarking that in some examples the makers left fingerprints and air bubbles (1975: 124). The differences in execution could have led to confusion in the dating, but Higgins' dating of the Tanagra figurines corresponds well with the context of those found at Rhitsona (1986: 82).

Stylistically, the manufacture of the figurines is more closely related to the potter's craft than that of the coroplast. The limbs and body are made of cylinders of clay, and the details are created through a combination of incision, addition, and hand-modelling. Furthermore, the shape and striped decoration of the monkeys bear a resemblance to handles on Geometric Boeotian amphorae (Preston 1975: 124). Higgins suggests that the monkeys may have come from the same workshop as contemporary Boeotian kylikes (1986: 82). Preston emphasizes, however, that even the handles of Geometric amphorae were created with more care than the figurines, and their significance is far from clear (1975: 126). Most have the long muzzle of baboons, and Higgins maintains that Thoth's baboon, which was probably introduced to Boeotia through Cypriot terracottas, was the most likely inspiration for these objects (1986: 82). Preston, however, suggests the figurines were probably toys or amusing household objects buried with the deceased, particularly because they are so poorly made. Nevertheless, he acknowledges the possibility that they may have served some religious or apotropaic funerary function (1975: 126).

It is important to recognize, though, that none of these monkey figurines have the ithyphallic characteristics either of Thoth's hamadryas or the later comical figurines that would also be produced in Boeotia. These monkeys are either mothers, females without infants, or devoid of gender altogether. The inspiration for the figurines, therefore, is more likely mother-monkey amulets and aryballoi than Thoth figurines. It certainly cannot be proven that these Boeotian monkeys had any apotropaic value, but they do not seem to have had comic value either. Their placement in graves suggests that they were valued by the owners, perhaps children or women, but insufficient reports of the contexts in which the objects were found warrant only cautious suggestions about the significance of the monkeys. It seems safe to say, however, that these Boeotian monkeys were not created with the same flippancy as the later examples that will be discussed in conjunction with Aesopic fables.

Karageorghis (1994) has also published a number of terracottas from Cyprus, dating from the Cypro-Archaic period, that are also worth mentioning here. Like the Boeotian terracottas, these were also found in tombs, and their identification as monkeys poses problems. Because most of the figurines are rather crudely made and all are highly stylized, some may depict bears rather than monkeys. These terracottas demonstrate a much greater degree of variation in form and quality than the Boeotian figurines discussed above. A comparison of several figurines may serve to highlight the variety. The first figure depicts a seated animal, probably a baboon judging by its elongated muzzle, which had round ears and stylized limbs (1994: 64-5, fig. 2). Painted stripes decorate its upper body and arms. Painted circles indicate the eyes, but the nostrils and mouth are incised. Its posture is similar to the Boeotian figurines in that its hands rest on its feet, creating a triangular form, and its sex is not indicated. The ears and extraordinarily long face are almost reminiscent of an aardvark. Two figurines from the Cesnola collection, now in the Metropolitan Museum of Art, depict animals with rounded heads, small round ears and elongated bodies. The first of these has a less refined appearance (**Fig. 85**; 1994: 64, 67, fig. 7). Its eyes were created from applied clay pellets, and other features are indicated with paint. The short legs are slightly drawn up, and the long right arm rest on the right knee. The left arm is raised but broken. The second has attractive facial features rendered in paint, and a raised nose also highlighted with a painted stripe. The body of this animal is more stylized than the previous example and is fully decorated with painted stripes. It holds an object in its hands, which Karageorghis identifies as a baby monkey (**Fig. 86**; 1994: 64, 67, fig. 8). I am not entirely convinced of this. The object is small and of an indistinct shape, so it is just as likely to be a piece of fruit or a toy. Another figurine from Aradippou, however, is definitely holding a baby. This figure has crude, pinched out features and painted decoration to indicate fingers. A continuous painted band also runs along the legs and lower back; another runs across the shoulders, beginning from circular bands painted on the upper arms. The baby's form is poorly defined, but arms, legs and a head can be distinguished (**Fig. 87**; 1994: 64, fig. 9). One figure, also crude and stylized, can clearly be identified as a baboon. The muzzle and brow-ridge are clearly rendered, and the animal holds its left hand to its mouth, as if eating a piece of fruit. The baboon is also ithyphallic and holds its penis with its right hand (**Fig. 88**; 1994: 68, fig. 10). Although this particular example seems to be a caricature of the Egyptian baboon, it does illustrate the continued relationship between the cultures. The final example, on the other hand, suggests a Near Eastern influence. A stylized, bearded male figure clasps straps to his chest, which support a small figure that clings to his back. Karageorghis acknowledges that the small figure has been interpreted as a human infant but insists that the 'infant' has simian features (**Fig. 89**; 1994: 71, fig. 14). If the little bundle is a monkey, the figurine is highly reminiscent of the Neo-Assyrian reliefs of tribute bearers, discussed in Chapter 2.

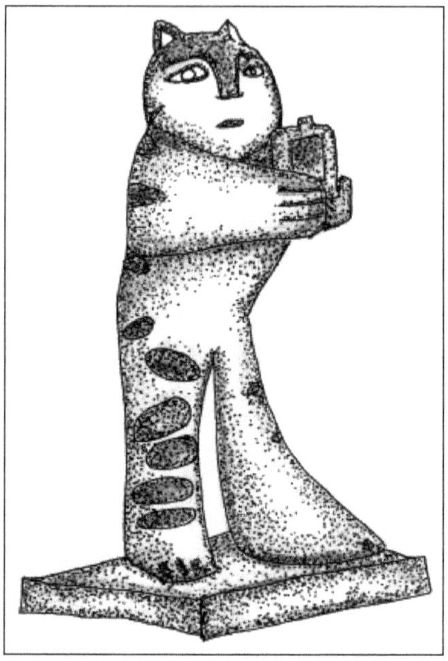

Fig. 86: A Cypriot monkey
(After Karageorghis 1994: fig. 8)

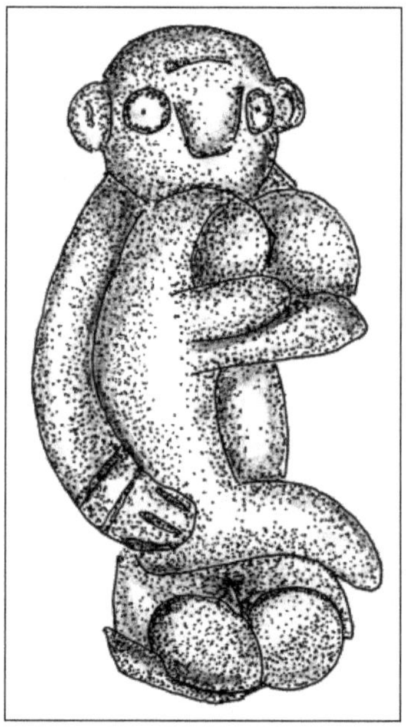

Fig. 87: A mother and baby monkey from Cyprus
(After Karageorghis 1994: fig. 9)

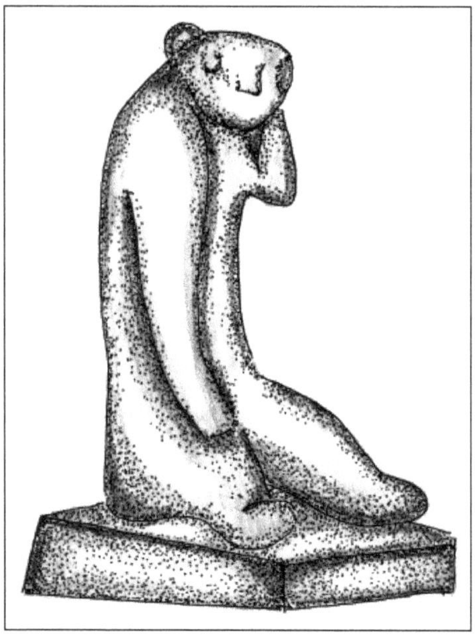

Fig. 85: A Cypriot monkey
(After Karageorghis 1994: fig. 7)

It seems probable that Egyptian and Phoenician prototypes were the most likely inspiration for these Cypriot monkey figurines and vessels, but the species most commonly

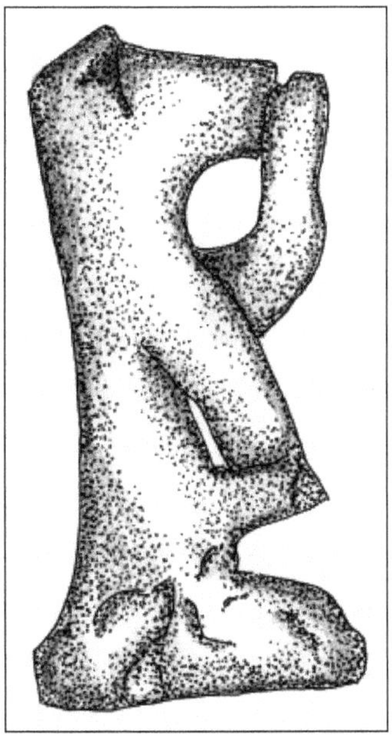

Fig. 88: A masturbating baboon from Cyprus
(After Karageorghis 1994: fig. 10)

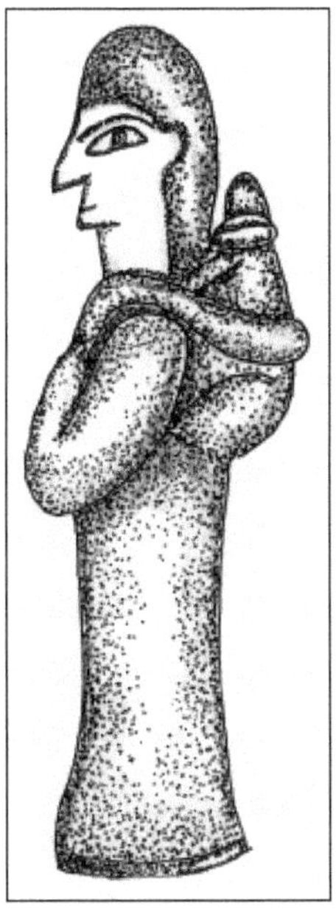

Fig. 89: A male figure carries either a baby or a monkey
(After Karageorghis 1994: fig. 14)

known in mainland Greece and Etruria would have been the Barbary macaque. These monkeys were imported from Phoenician sites in western North African, such as Mauretania Tingitana in modern-day Morocco and Carthage in Tunisia, so the depiction of the animals in North Africa and Etruria is worthy of consideration. Bonacelli catalogued a number of representations of the animals at Etruscan sites in the form of tomb paintings, figurines and decorative elements in metal work and suggested a Punic connection, mainly because the simians were represented without tails. His argument also has a linguistic basis—in Greek, the common word for monkey is *pithekos*, rather than *cercopithecu*s, suggesting that the monkey known in Greece and Italy was the tailless Barbary (1932: 343). He

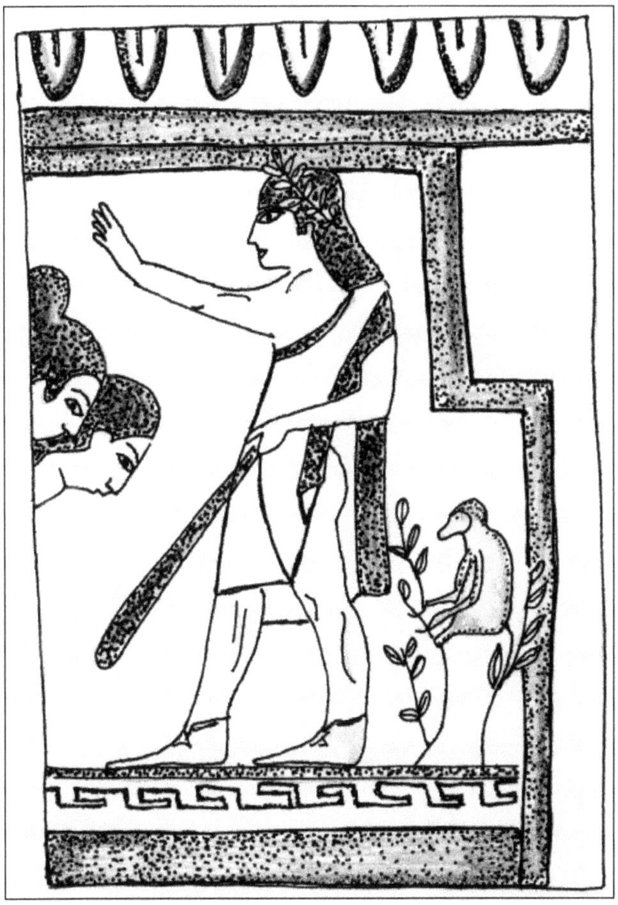

Fig. 90: Scene from the Tomba della Simia at Chiusi
(After Bonacelli 1932: pl. 16, fig. 2)

also argues that the Etruscan word for monkey, *arimi*, was borrowed from the Semitic *harim*, meaning a flat-nosed person. The word is found in the name of the island Inarime, the Greek Pithecussae, where the legendary monkey-like Cercopes were said have lived (1932: 346).[24] Bonacelli suggests that the word is Punic, but Rebuffet-Emmanuel maintains that the toponym could have arisen from any Phoenician exploration of the western Mediterranean (1967: 640).

At least two Etruscan tombs depict tailless pet monkeys. The Tomba della Simia at Chiusi, probably dating from the early to the mid 5th century BCE, contains scenes of athletic contests. In addition to the depiction of victorious athletes and their relevant games, there are comic figures, such as boys playing with dogs, dwarfs and a monkey, which squats on a stump behind a man crowned with myrtle. The little simian wears a collar attached to a large chain to keep it from straying (**Fig. 90**; Bonacelli 1932: pl. 16, fig. 2). In the 4th century BCE tomb known as Tomba Golini or the Tomb of the Vetii near Orvieto, another tailless monkey can be seen climbing a pole. The scene is fragmentary, but the monkey has a red cord attached to its ankle, which is held by a human hand (**Fig. 91**). McDermott remarks that because the tomb contains scenes of the underworld, it has been suggested that the monkey has religious significance and that the pole symbolizes a funeral stele. He suggests instead that the monkey is a pet of one of the people buried

[24] The transformation of the Cercopes into monkeys is discussed below.

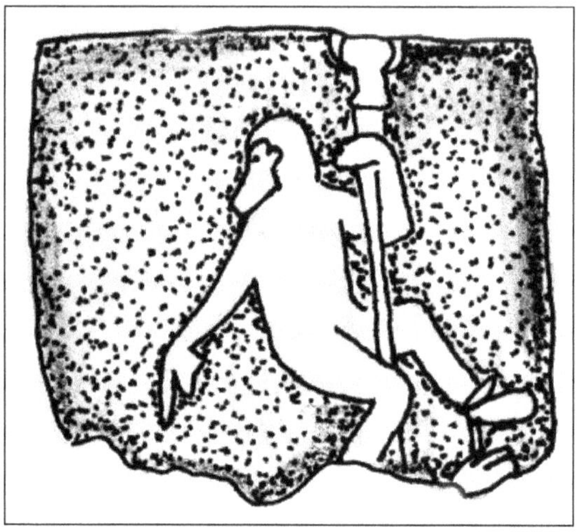

Fig. 91: Scene from Tomba Golini
(After Bonacelli 1932: pl. 16, fig. 1)

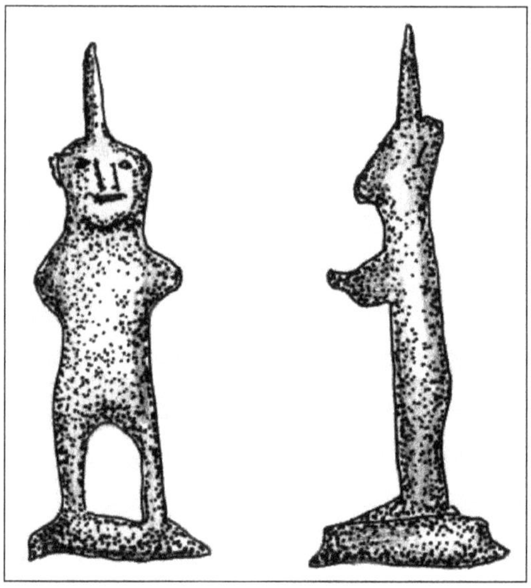

Fig. 92: A bronze monkey from Mauritania Tingitana
(After Rebuffet-Emmanuel 1967: pl 132, fig. a)

there, possibly the boy, Vel Leinies, who is also pictured in the tomb (McDermott 1938: 277-78).

It is important to recognize that, although the monkeys in these paintings are not instantly recognizable as a particular species, their lack of tail and their size in relation to the human figures in the scenes make the Barbary the most likely candidate. It seems reasonable to suggest that the Etruscan knowledge of this animal came from North Africa, and trade with the region was probably well established. Bonacelli suggests that amulets, dating from the 7th century BCE, found in Etruscan tombs were probably made in western North Africa and even argues that the same region supplied the amber and ivory from which the amulets were produced (Bonacelli 1932: 362). This provenance, however, is debatable, since these raw materials have also been found in Etruria. The amulets, therefore, could have been manufactured locally from imported ivory and amber (Rebuffet-Emmanuel 1967: 640).

Rebuffet-Emmanuel, in her 1967 article, "Singes de Mauritanie Tingitanie et d'Italie," found even more similarities between North African and Etruscan representations of monkeys. Her study mainly concerns a group of Punic simianesque bronze figurines from Morocco, dating from the 1st century BCE, that bear a striking similarity to earlier Syrian amulets of gods. Figurines of divinities, dating from the 2nd millennium, found at Minet el-Beida and Megiddo wear a bonnet or tall hairstyle, while standing monkeys from Mauritania Tingitana have a peculiar sword-like projection rising from the tops of their heads (**Fig. 92**). She argues that this mutation is not simply the product of a degeneration of the image but a sign of continued contact with eastern cultures that resulted in an intentional merging of the simian and the human/divine form. She maintains that, at the very least, the figurines have an apotropaic value but suggests they may also depict gods (1967: 644). Several earlier Etruscan bronze objects also incorporate figures of standing simianesque beings, and it seems reasonable to suggest a certain amount of cultural transference between Etruria, and eastern and western Phoenician sites. The Etruscan tomb paintings suggest that monkeys had value as pets, but the earlier amber and ivory amulets suggest an apotropaic value. These functions were probably adopted from Punic North Africa along with real Barbary macaques, and the Etruscan depiction of standing monkey may, in turn, have influenced the Punic merging of god and monkey.

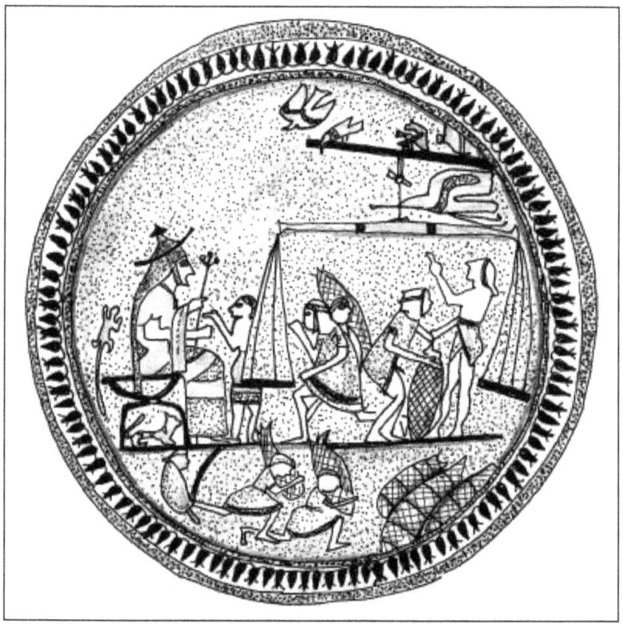

Fig. 93: The Arcesilas Kylix

A combination of eastern and western influence is also present in vase paintings. The Arcesilas kylix, which belongs to a group of black-figured Laconian pottery, dating from the 6th century BCE depicts King Arcesilas of Cyrene presiding over the weighing and loading of silphium on a ship (**Fig. 93**). The exotic nature of the scene is emphasized by depiction of animals, such as the big cat, possibly a panther, that sits under the king's chair, the lizard that seems to be suspended in mid-air behind him, and the monkey that sits on the rigging at the top right-hand side of the scene. McDermott comments that the monkey's collar identifies it as a pet (McDermott 1938: 220-21, no. 311). The exact species is not obvious. It has the long muzzle of a baboon but seems to be tailless. The placement of the monkey in the scene suggests to me that the artist may have been drawing upon the function of the Egyptian baboon. The scale upon which the silphium is weighed is suspended from the rigging just to the left of the monkey. Like Thoth-Baboon in the vignettes from Spell 125 from the Book of the Dead, the monkey sits above the balance, as if ensuring the fairness of the weighing. The monkey, then, may have functioned both as a valuable exotic pet and as a symbol of fair business.

Another Laconian kylix, excavated near the Temple of Artemis Orthia in Sparta, incorporates a monkey as a decorative element in the scene inside the cup (**Fig. 94**; Droop 1929: 85, pl. 9). The main figures are four bearded men, who run along the border of the scene. They look

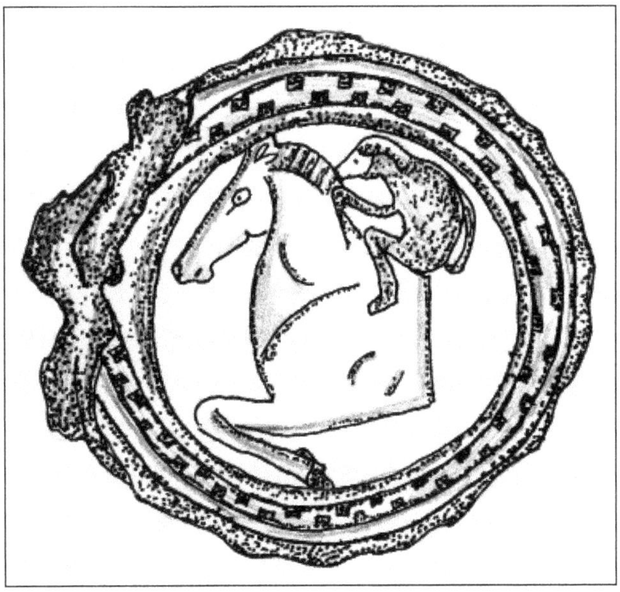

Fig. 95: A monkey riding a horse (After McDermott 1938: pl. 3)

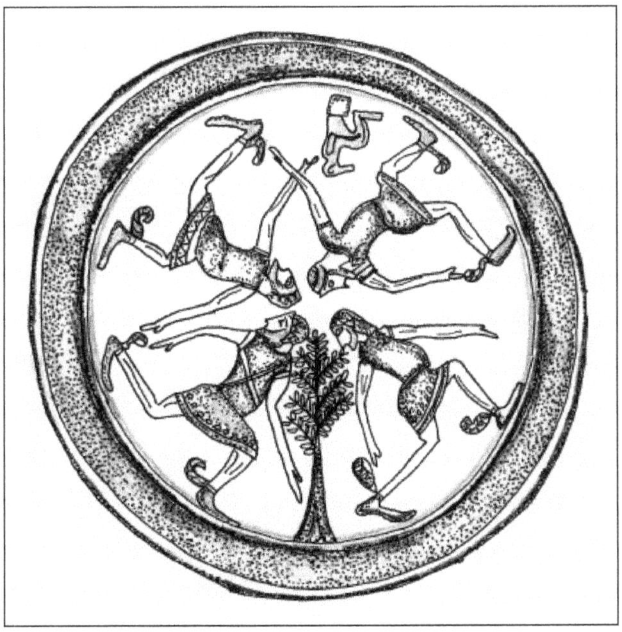

Fig. 94: Laconian kylix with monkey (After Droop 1929: pl. 9)

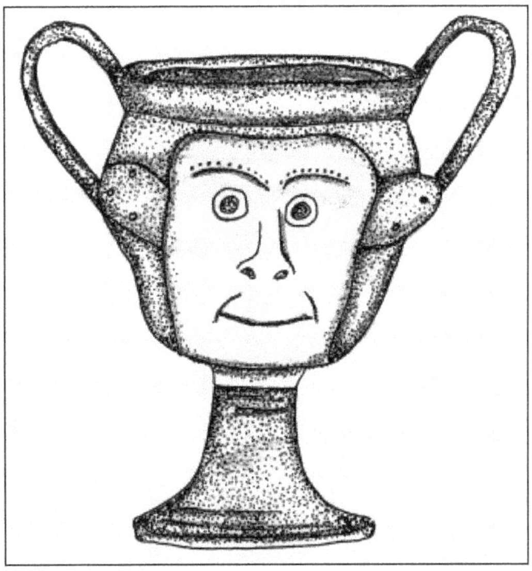

Fig. 96: Terracotta kantharos, Cesnola Collection, Metropolitan Museum of Art (74.51.369)

back over their shoulders, and their heads almost meet at the centre of the scene. The men have wings attached to their ankles, and they are divided into pairs by the insertion of a small tree rising up from the border and by a monkey sitting on a stump that rests on the border directly opposite the tree. McDermott suggest the men represent the four winds, while the monkey, which is rather crudely painted, is a burlesque, unconnected to the rest of the scene (1938: 222, no. 312). He is probably correct in this assertion. Although the monkey resembles a baboon because of its long, square muzzle, it does not seem to possess any religious or apotropaic significance. A similar use of the monkey can also be found on the famous Protocorinthian Macmillan aryballos. The third register from the base depicts hunters on horseback. A swan sits between the legs of one of the galloping horses, while under the horse to the right, a stylized crouching monkey raises its fist. The little creature may have been inserted as a humorous element to an otherwise serious scene. Another vessel with painted decoration that connects monkeys with horses is a black-figured kylix in Athens National Museum. The inside scene depicts a furry little monkey sitting riding on the back of a horse and grabbing its mane for support (**Fig. 95**; McDermott 1938: 224, no. 316).

One truly unusual vessel from the Cesnola collection is a kantharos in the form of a monkey's head (**Fig. 96**). One side of the vessel is beautifully modelled into the face of a monkey, and its features are carefully emphasized with black painted decoration. The rounded, protruding ears each have three perforations. The kantharos dates from the 7th century and was probably created by a Cypro-Phoenician potter (McDermott 1938: 240, no. 335).

Despite these mainly positive representations of monkeys, the animals did not fare particularly well in literature. The first Greek citation of an Aesopic animal fable comes from Hesiod's *Works and Days* 202-212, in which a hawk catches a nightingale and tells her not to resist because only a fool tries to withstand the stronger (Perry *Aes*. 4a). Fables involving monkeys, however, are first mentioned in fragments from an epode of Archilochus, writing in the mid-seventh century. There, the monkey is the dupe of the fox (Frags. 224, 233 in Laserre 1959; Perry *Aes*. 81). Before explaining the monkey's role in fable, though, a brief explanation of fables in Greco-Roman literature is in order. Originally, the tales were used as exempla in poetry and prose, as was the case in Archilochus' epode, which seems to have been addressed to a friend called Pericles (Lasserre 1959: 65). At this early date, no corpus of fables existed, and they were not considered a literary genre, as such. The first extant collection was produced in Latin by Phaedrus, a freedman of Augustus, sometime in the first century CE. Slightly later, Babrius, a resident of Syria, produced a collection of fables in Greek. Their literary value was based solely on the fact they were written in verse.

Aesop, if he existed at all, would have lived in the sixth century BCE, and stories surrounding the events of his life contain more legend than fact. He was believed to have been a slave from Thrace or Phrygia, who was born mute. Isis granted him the power of speech for his piety, and the muses granted him eloquence. He used his rhetorical skills for the benefit of his owner, Xanthus, who eventually freed him. Aesop then travelled to the court of Croesus and had a series of adventures. Eventually, he made his way to Delphi, where he was unjustly killed, according to some legends because one of his fables insulted the local people. Physically, Aesop was ugly and deformed, and a few vase paintings and figurines of a lame, balding or hunchbacked man are thought to be representations of the fabulist. It is important to note that even ancient authors did not believe Aesop invented fables. He was simply credited with using them in the most beneficial and creative ways.[25] References to the so-called Aesopic fables have been found in a variety of earlier literature from the Near East and Egypt, including the Old Testament, i.e. 2 Kings 14.9. As mentioned in my first chapter, Thoth even persuaded Hathor-Tefnut to return to Re using fables that were later attributed to Aesop. Clearly, then, fables were imports to Greece, as were a number of the animal characters in them, including monkeys.

Over a dozen extant fables revolve around the antics of these exotic creatures, and they are arguably the greatest indicator of attitudes toward monkeys from the Archaic Period onward. Like Rodney Dangerfield, the poor creatures get no respect, and like the tales concerning Aesop, a number of fables outline the extreme ugliness of monkeys. Babrius (and much later Avianus, writing in the late 4th or early 5th century CE) tells the story of a beauty contest for babies organized by Zeus. All the animals bring their offspring to the god to be judged, including the monkey, who clutches her snub-nosed, naked baby to her breast. All the animals laugh at the sight, but the monkey's mother replies, "Zeus knows who the winner is; this one is the most beautiful of all, in my opinion!" The moral of the story, provided in the epimythium, is that each person believes his own child to be attractive (Perry 1952: 364). Another fable in the category of ugly simians mocks the animal's bare bottom and lack of tail. (The monkey in question is must be a Barbary macaque). The self-conscious monkey asks the fox for part of his tail to cover his bare behind. The nasty fox refuses the request with the added insult that even if his tail were longer, he would drag it through the mud rather than share a portion of it with the monkey (Perry 1952: 533). Phaedrus also tells an ugly monkey joke in the form of a fable: A man sees a monkey hanging in a butcher's shop with meat and other items and asks what its taste is like. The butcher replies that it tastes as bad as it looks. Phaedrus explains in the epimythium that this must have been told in jest, since he had come across many attractive people who were horrible and many excellent people who were hideous (Perry 1952: 496). McDermott emphasizes that the monkey in the fable was a pet rather than a source of food—it was probably hanging in a cage rather than on a meat hook—and that the word *sapere* is a pun concerning its taste or knowledge (1938: 114-15). Toynbee also describes a marble relief found on the Via della Foce at Ostia, dating from the second century CE, which adds iconographical proof that monkeys were only kept in shops as amusing pets. Two salespeople, a woman and perhaps a young man, stand behind the counter and interact with the customers. In the background to the left of the salespeople, ducks are hanging upside down on meat hooks. To the far right, two monkeys are sitting on top of rabbit cages. As Toynbee remarks, the simians are not tied up and seem to be pets put on display in the shop for the customers' enjoyment (**Fig. 97**; 1973: 57, fig. 13). If the customers in the shop find the animals ugly, the monkeys do not seem to be aware of it. One smiles gently, while the other just scratches its head.

Fables also ridicule the monkey's inclination to imitate humans, whether it understands a particular action or not. One fable tells the woeful tale of a monkey who watched some fishermen casting a net in the river and retrieving a load of fish. As soon as the men leave the area to eat lunch, the monkey climbs down from his tree and tries to catch fish with the net. Instead, he becomes entangled in the net and drowns. With his last breath, the monkey admits that his fate was deserved because he should never have tried to practice a skill that he had not learned (Perry 1952: 203). In another fable, the monkey's attempt to impersonate his master also results in tragedy. A man brings his pet monkey on a ship that subsequently sinks into the sea. A dolphin sees the monkey drowning and gives him a ride on his back, thinking that he has rescued a man. The monkey plays along, but ultimately his deception is revealed as the dolphin reaches the port of Athens. When asked if he knows the Piraeus, the monkey responds that Piraeus is a very close friend of his. The dolphin realizes he has been tricked and drowns the monkey (Perry 1952: 73).

[25] Discussions of Aesop and other authors of Aesopic fables can be found in Holzberg (2002) and Perry's introduction to the Loeb translation of Babrius and Phaedrus (1965). Perry's *Aesopica*, published in 1952, is perhaps the most complete and user-friendly compendium of all known fables and references to fables in Greek and Latin literature. (In my citations of this work, the number refers to the fable number rather than the page). More recently Adrados has attempted to reconstruct the history of fables in a considerably less user-friendly three volume set, entitled *Historia de la fábula greco-latina* (1979-1987).

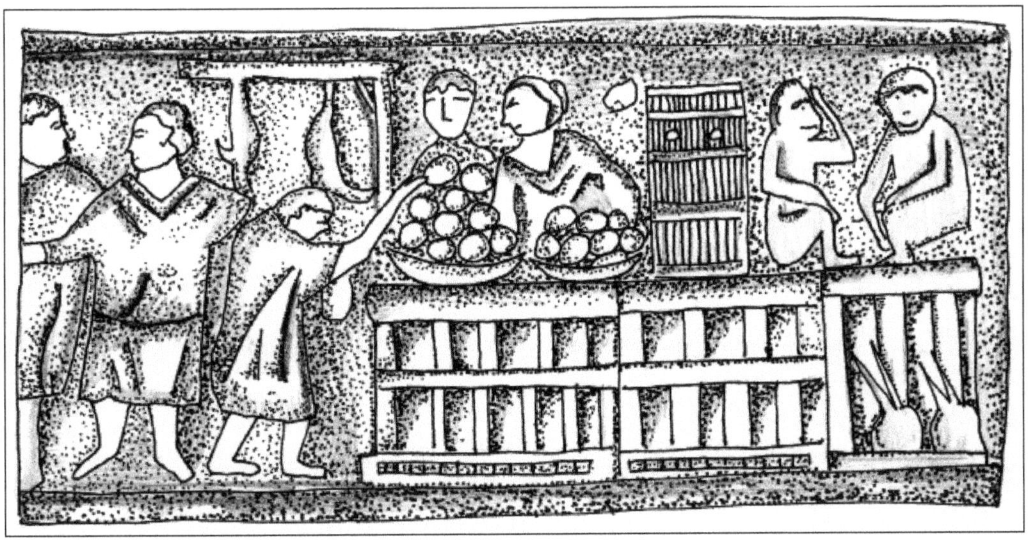

Fig. 97: Marble relief from Via della Ostia (After Toynbee 1973: fig. 13)

Less cruel fables relate to the monkey's comic ability to imitate dance moves. Lucian tells one of the best known: An Egyptian king once taught some monkeys the Pyrrhic dance. The animals were quick learners and even wore masks and purple robes. Their dance was a great hit with the audience, until one of the spectators threw nuts at them during a routine. The monkeys forgot all about the dance and ripped off their costumes to grab for the nuts. Then, of course, they were laughed off-stage by the audience (*Piscat.* 36; Perry 1952: 463). McDermott noticed that the story also seems to have led to a Greek proverb about a monkey in human finery that was similar in meaning to the English saying: An ape's an ape, a varlet's a varlet, though they be clad in silk or scarlet (1938: 117). The pagan emperor Julian, upon returning from a campaign in Gaul, was mockingly called *purpurata simia*, apparently in reference to this fable (Ammianus Marcellianus, 17.11.1). From time to time, though, it is the monkey who is imitated to the detriment of other animals. At a gathering of animals, the monkey created a sensation with his dance. When the camel tried to imitate his moves, the other animals were disgusted at the pathetic attempt and drove him from the party (Perry 1952: 83). Babrius tells the fable of a donkey that climbs the roof and breaks some tiles. When he is beaten for the damage he caused, the donkey complains that the monkey was praised for jumping on the roof only the day before (Perry 1952: 359).

A number of fables make the fox the clever adversary of the foolish monkey. When the monkey boasts to a fox that the stele before them was erected in honour of his grandfather, the fox tells the monkey to lie as he wishes, since the truth cannot be proven (Perry 1952: 14; Babrius 81 in Perry 1965). This story also suggests that monkeys are boastful liars, which also becomes a theme in Latin literature. Several of Plautus' deceitful or ridiculous characters, for example, have names related to *pithekos* or *simia*, such as the servant girl, Pithecium, in the *Miles Gloriosus* or Simia in *Pseudolus*, whose name derives from the Greek, Σιμίας. The play on words, however, is clearly intentional (McDermott 1936: 152). At another assembly of animals, the monkey again makes such a good impression that he is elected king. The fox becomes jealous and tricks the monkey into taking some meat from a trap. When the monkey is caught and complains that he was duped, the fox asks how he could presume to be king with brains such as his (Perry 1952: 81). In another fable, the monkey gets the better of both the fox and the wolf. The latter brought a charge of theft against the fox, who pleaded not guilty. The monkey listened to both arguments and declared that the wolf had not lost the item he sought and that the fox was guilty of stealing it just the same. Phaedrus states in the promythium that Aesop's fable demonstrates how someone who has acquired a reputation for deceit will not be believed, even when he tells the truth (Phaedrus I, 10 in Perry 1965).

Monkeys also play a variety of other roles in fables that incorporate both their positive and negative qualities. In one, a shop keeper has a monkey guardian whose skills in preventing theft were said to be unsurpassed. A friend of the shop keeper asks to test the monkey's abilities by trying to steal something. When the man enters the shop, he tricks the monkey by playing upon its inclination to imitate human actions. He made a number of gestures and expressions that culminated in pressing his eyes closed with his fingers. When the monkey made the same gesture, the man seized the opportunity to walk off with an item from the store. The monkey was then beaten by the shop keeper so that he would realize his mistake. When the man attempted the same trick a second time, the monkey pulled his eyes wide open to demonstrate that he could only be fooled once (Perry 1952: 643). Finally, the monkey's maternal instincts are called into question in a fable told by Babrius (Perry 1952: 218). A monkey who gives birth to twins loves one baby and despises the other. The one she loves is coddled to death, while the hated son is cast aside and survives. Babrius' epimythium warns that this is the custom of many people and that they are a type to be despised. In Avianus' version of the same fable, the mother is forced to drop her favoured son while fleeing from a predator. He dies, but the other holds tightly to her back and survives. In time, she grows to love the surviving infant as much as her first choice. Avianus' epimythium is a little more philosophical: Many people come to appreciate that which they have neglected, and the humble may have hope of a better life (Avianus 35).

The Aesopic fables clearly reflect the ambivalent attitudes the Greeks and Romans were developing toward monkeys from at least 7th century BCE. Other literary evidence brings together some of the recurring themes. A fragmentary passage from Semonides of Amorgus, writing in the 7th century BCE, defines different types of women and names the animals from which Zeus created them. All are wicked except those who take after the bee. Those who take after monkeys are the very worst:

> ....this is the biggest plague of all that Zeus has given to men. Her face is hideous....She is short in the neck; she moves awkwardly; she has no bottom, and is all legs.... She knows every trick and twist, just like a monkey; she does not mind being laughed at, and will do no one a good turn, but...spends the whole day planning, how she can do someone the worst possible harm (Lloyd-Jones, Trans. 1975: 50).

Here again, the poor monkey is considered ugly, ridiculous and dishonest—a nasty piece of work altogether. Although the much later author, Aelian, did not despise monkeys, as discussed further on in the chapter, he did insist that monkeys could be extremely cruel creatures, particularly when they tried to imitate human actions. He relates the tale of a monkey who, from a distance, watched a nurse bathing an infant in a tub. When she took the baby out of the tub and dried him off, the monkey watched where she put him. As soon as she left, the monkey jumped into the room, put the baby back in the tub, and poured a nearby pot of boiling water over him. The child, of course, died from his burns (*NA*.7.21).

It also seems that these Hellenized tales began to spread east again and may have influenced Indian fables. In a presentation on the collection of Indian fables known as the *Pañcatantra*, Thite cites a number of fables involving monkeys (1983: 46), which bear strong similarities to the Aesopic fables. Although Thite does not conclude that these particular fables were influenced by the Greeks, Adrados remarks in the subsequent discussion that both Greek and Mesopotamian fables had a decisive influence on Indian fable from as early as the third century BCE (discussion in Adrados and Reverdin 1983: 54). In one of the fables cited, a monkey outwits a crocodile. The crocodile allows the monkey to ride on its back on condition that she may eat its heart afterwards. The clever monkey claims that his heart is in the tree where he lives and that he will return with it. Of course, the monkey jumps to the safety of his tree and never returns to the crocodile. A similar motif of trickery is found in a number of Aesopic fables, but there it is usually the wolf who is duped by his prey. Another monkey tale cited by Thite, in which a monkey dies pulling out a wedge, an activity that should not concern him, is very reminiscent of the fable of the monkey and the fishermen (1983: 40).

In Greece, though, the fables may also have inspired works of art. Lissarrague makes the interesting suggestion that a connection exists between Aesopic fables and certain terracotta figurines. He maintains that some Boeotian and Corinthian terracottas, dating from the 6th-5th century BCE, anthropomorphize the animal world and reflect Aesopic themes, particularly the monkey and her children and the monkey and the fox (Lissarrague 2000: 140). He selects a number of thematic groups from Winter's often cited catalogue of figurines, *Die Typen der figürlichen Terrakotten*, published in 1903. In trying to find an artistic representation of the monkey and her children, Lissarrague makes the rather peculiar choice of a Boeotian smiling male monkey supporting a younger monkey on his shoulder (Lissarrague 2000: 141, fig. 5.5). It seems highly improbable to me that this particular example was influenced by any known fable. The monkey holding a fox, on the other hand, is somewhat more convincing (2000: 141, fig. 5.6). He notes that the image does not reflect any situation related in fables but maintains that it shares the spirit of the stories. Indeed, any artistic rendition of a story would have to be presented in an abridged form, and it seems entirely logical that the monkey's presence in fables inspired artists to create these comical figurines. It is also reasonable to accept Lissarrague's suggestion that terracottas depicting monkeys sitting on turtles and pigs or holding birds are meant to parody the human world and cause laughter (2000: 141-2, figs. 5.7, 5.8., 5.9). However, because pigs, turtles and birds do not appear with the monkey in fables, it is difficult to argue that fables are the ultimate inspiration for the figurines. Mitten, however, has suggested that the motif of a man or monkey riding a tortoise may indeed be reminiscent of a fable concerning the frustration of choosing a slow horse. He proposes alternatively that such figurines may parody an epic theme. He notes that one of the metopes of the Sele Heraion depicts a male figure riding a turtle and that scholars have variously interpreted the figure as Odysseus being rescued from the Charybdis or as Tantalus (in Kozloff 1981: 143).

This line of reasoning may be strengthened by the fact that monkeys have been used elsewhere, particularly in later Roman art, to parody mythological heroes. One of the best known examples comes from a painting found at Pompeii now in the National Museum in Naples, depicting the flight of Aeneas from Troy with his son and father. All three are depicted as baboons with long tails. Both wear short cloaks, and Ascanius wears a Phrygian cap. Anchises, carried on the left shoulder of Aeneas, is nude and holds a dice box in his right hand. The painting seems to make an anti-imperialist statement against the self-proclaimed descendants of Iulus, even mocking Julius Caesar's penchant for gambling through Anchises' dice box. McDermott remarks that the painting also attacks Virgil. The *Aeneid* came under attack as an imitation of Homeric epic, and the simian caricature indicates that Virgil was "aping" Homer (1938: 279-80). Additionally, a Roman lamp in the Louvre (4922) depicts a monkey being carried off by an eagle. This monkey also wears a Phrygian cap, suggesting a parody of Ganymede being abducted by Zeus (McDermott 1938: no. 561, Pl. 10). This lamp may, in turn, find a parallel in an episode from the Golden Ass. As he seeks the priest of Isis who will be his salvation, Lucius, the hero-turned-donkey, describes the festive procession in which a monkey sports a Phrygian cap and carries a golden cup, in imitation of Ganymede (11.8).

In addition to their use in fables and parodies, monkeys had a connection to satyrs and comic actors. Brijder cites two

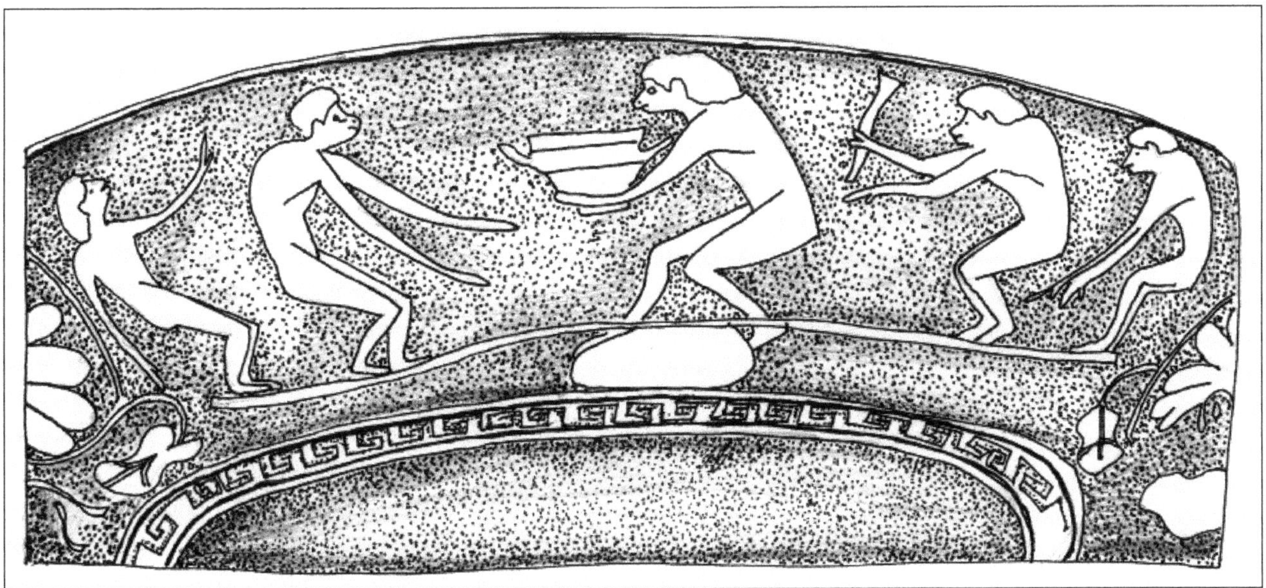

Fig. 98: Kylix depicting actors in monkeys masks (After Brijder 1988: 62, fig. 1)

examples of nude comic actors wearing monkey masks. On a scene on the outer side of a kylix found in the Tomba della Panathenaica at Vulci (Vulci 64224), five nude men stand on a see-saw (**Fig. 98**). All have short, bowl-shaped hair crowned with ivy- wreaths, and all are wearing monkey masks. The centre actor faces left, holding a large drinking bowl.

The actor facing him seems to be reaching out for the bowl, while the actor to the far left seems about to fall off the balance beam. Behind the centre actor are two others; the closest holds a drinking horn. The kylix dates from c.520 BCE. The second scene, from an askos in the British Museum (E 740), depicts two actors walking around the top of the vessel on all fours. Again, one of the actors is holding a drinking horn (Brijder 1988: 62-3, figs. 1, 2). He suggests that certain figurines, typically identified as comasts or satyrs, may also represent comic actors disguised in monkey costumes. An aryballos in the Allard Pierson Museum (3402), for example, is in the form of a man with long hair and a beard squatting like a monkey (**Fig. 99**). His animalistic qualities are emphasized by the spots on his body and his enormous pot-belly. His navel is indented and his genitalia are pronounced (1988: 67, fig. 7). He may be compared to a squatting monkey aryballos in the Museum of Fine Arts, Boston (**Fig. 100**). The little monkey sits with his knees drawn up to his chest, which gives him an appearance that is almost as rounded as the satyr (Fairbanks 1928: 175, no. 502; Richter 1930: pl. 58). The blurred line between monkeys and satyrs has also been recognized by Lissarrague, who compares the characteristics ascribed to both in his article, “L’homme, le singe et le satyre” (1997). Satyrs, which are frequently represented on pottery of 5th-4th centuries BCE, demonstrate a ‘quasi-humanity’ that places them in an imaginary realm between the human and animal. Representations of their bodies, sexuality, and general behaviour lend themselves to parody, and it seems that monkeys were used in a parallel fashion, albeit with less frequency.

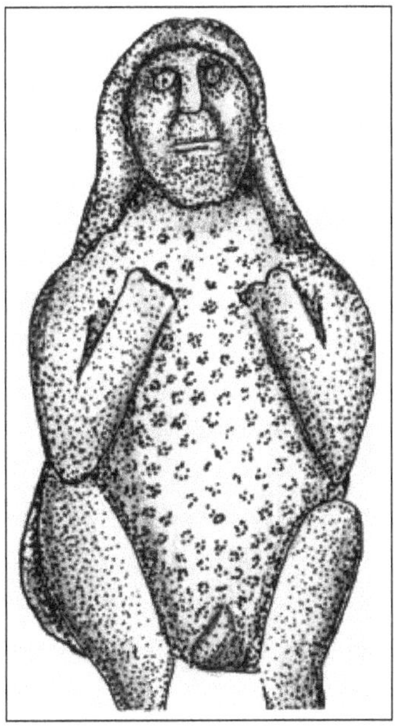

Fig. 99: Aryballos in the Allard Pierson Museum (After Brijder 1988: 67, fig. 7)

Satyrs are generally depicted with human bodies and equine tails and ears. Occasionally, they have hooves or fur on their bodies. Their animal nature is further illustrated by their behaviour. Their wildness makes them suitable attendants of Dionysus, and they form the male counterpart to Barbarian women and maenads (Lissarrague 1993: 207). Many of their characteristics typify behaviour inspired by the god. In tradition and iconography, they are associated with musical entertainment and dance. The story of Marsyas’ musical competition with Apollo is well known and frequently depicted, and numerous painted vases represent satyrs dancing and playing musical instruments. Their fondness of wine is another commonly depicted trait, which is reinforced by their

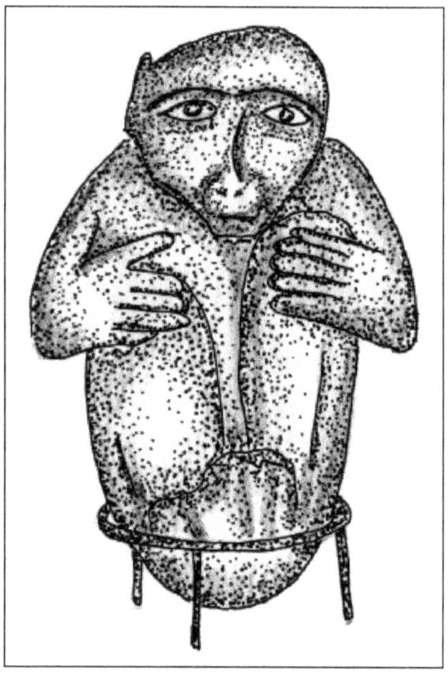

Fig. 100: Monkey aryballos in the Museum of Fine Arts Boston (After Richter 1930: pl. 57, fig. 185)

frequent appearance on drinking cups and kraters. This drinking is often combined with their sexuality, as on a red-figure psykter in the British Museum, E 768. In one scene, a squatting satyr leans backward, supporting his upper body with his arms. His huge, erect penis supports a kantharos. Another satyr to the right approaches with an oinochoe to fill it, while a third satyr approaches to the left with another kantharos (Lissarrague 1990: 72, Fig. 2.8). The satyr's enormous phallus is in opposition to the Greek ideal of male beauty. Attractive youths are invariably portrayed with small genitalia, and the literary evidence supports the iconography. In Aristophanes' *Clouds* (1011-14), Dikaios Logos describes the ideal ephebe (personal translation):

> ἕξεις ἀεὶ στῆθος λιπαρὸν,
> χροιὰν λευκὴν, ὤμους μεγάλην,
> γλῶτταν βαιὰν, πυγὴν μεγάλην,
> πόσθην μικράν.
>
> You will have a sleek chest,
> pale skin, broad shoulders,
> a short tongue, big bottom,
> and a small penis.

As Lissarrague argues, only old men and barbarians are depicted with large penises, so the satyr's enormous genitalia and aroused state should not be viewed as signs of hypervirility. Instead, these characteristics serve to place the mythical creatures securely in the animal realm and inspire ridicule (1990: 56). Sexually, satyrs are often portrayed masturbating or having relations with donkeys or deer. When satyrs masturbate, they are generally depicted crouching, a position that is associated with defecation and servitude. Additionally, their occasional frontal depiction enhances the obscene quality of their posture and endowments (1990: 56-7). This is evident from the depiction of a satyr on a black-figured amphora in Berlin, 1671 (Lissarrague 1990: 70, figs. 2.4-2.5). As a general rule, satyrs are not shown engaging in full sexual intercourse with women. More commonly, they pursue or carry off maenads, and scenes of mutual exhibitionism are not uncommon (1990: 62-3).

Like satyrs, monkeys do not match the Greek ideal of inner or outer beauty. As the passage from Semonides demonstrates, monkeys were considered ugly because of their lack of rump and short necks. Thersites, portrayed in the *Illiad*, 2.217-219, as an ugly trouble-maker willing to say anything to get a laugh, is mentioned in Plato's *Republic*, 10.620c, as someone who will be reincarnated as a monkey. McDermott also remarks that in a passage from Sophocles' play, *The Trackers*, 121-122, the satyrs are likened to monkeys because of their stooping postures and shouting (1935: 173). In plastic arts, Lissarrague remarks, the monkey's personality traits, such as flattery and braggadocio, are difficult to render, and the artists tended to portray only those characteristics that could be physically demonstrated: ugliness, mobility, agility, and mimetic gesture (1997: 462). All of these characteristics could equally be applied to representations of satyrs and human actors in monkey costumes. The monkey's agility and fondness for wine are implied in the gestures of the actors depicted on the kylix and askos, described above. Although Aelian states that all animals avoid wine, he insists that some animals are more prone to its effects than others (*VA*.2.40, personal translation):

> If the monkey and the elephant drink wine, the elephant forgets its strength, and the other its trickery; and they are readily captured.

Monkeys also share the satyr's association with music. Two terracottas from Cyrene, one in the Louvre and the other in the British Museum, apparently made from the same mould, dating from the 3rd century BCE, depict a monkey, seated on a rectangular base, playing a trigonon and wearing a diadem (**Fig. 101**; Lissarrague 1997: 464, fig. 13; Mollard-Besques 1986: no. D4395; McDermott 1938: no. 102; Winter 1903: 225, fig. 9). Another figurine group from eastern Greece also depicts a seated monkey holding a lyre in its left hand (Winter 1903: 222, no. 9). These examples do not come from mainland Greece and were probably inspired by Phoenician/Punic or Egyptian prototypes. The monkey dancers in Lucian's tale were also trained by an Egyptian king, and Aelian tells of baboons in Ptolemaic Egypt that were taught to recognize writing, dance, and play musical instruments. After performing their tricks, the baboons would then collect payment from the spectators, which they popped into their bags (*NA* 6.10). This ability to entertain through an imitation of human actions also links monkeys to comic actors. A Roman bronze figurine of an actor wearing a baboon mask attests to the continuity of the association across cultures (Lissarrague 1997: 469, fig. 25).

Additionally, the crouching posture of satyrs is reminiscent of the Egyptian portrayals of Thoth as a baboon, which are almost always ithyphallic. It is easy to imagine how such representations would have been ridiculed by the Greeks, who considered the posture a sign of low status and the large phallus as a sign of animalistic barbarity. The monkey's connection with satyrs seems inevitable, and a

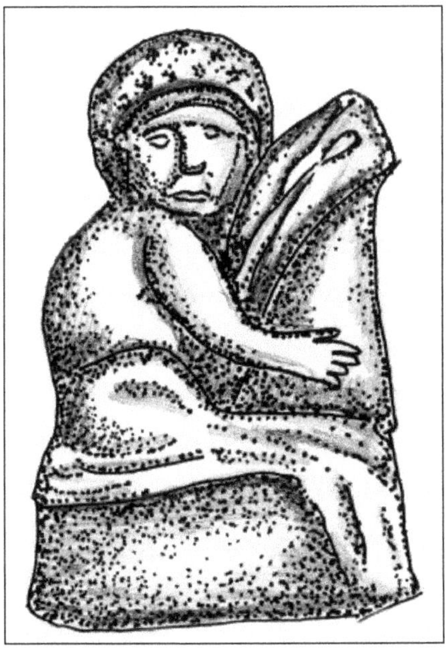

Fig. 101: Monkey playing a Trigonon (After Lissarrague 1997: pl. 12, no. 13)

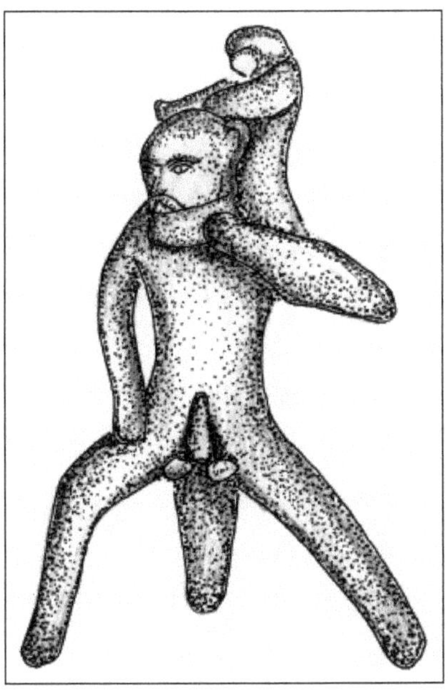

Fig. 102: A satyr with a monkey on his back (Goulandris Museum)

terracotta figurine in the Goulandris Museum suggests that the two creatures are close companions. A nude ithyphallic satyr stands with his legs wide apart and uses his left arm to support a monkey that clings to his head (**Fig. 102**). Interestingly, though, monkeys were rarely portrayed masturbating or engaged in sexual intercourse.[26] Lissarrague argues that monkeys might have been considered a sort of 'sexual zero' on the basis of an episode from Lucian's *Lucius or The Ass*, in which the hero, having recovered his human form, returns to his lover, hoping that she will find him even more attractive as a man. However, when she catches sight of his considerably smaller member, she exclaims (56, personal translation):

> I thought that you would still have that gorgeous jewel that distinguished my donkey, but I see that, from this beautiful and useful animal, you have transformed into a monkey.

The baboon's aggressive sexuality, on the other hand, was acknowledged in literature. Aelian maintained that the animals were lascivious and attacked women and children (*NA* 7.19).

Hoffman also recognized that terracotta monkeys are sometimes depicted imitating gestures associated with satyrs and suggests that they may be considered "satyrs transformed into monkeys" (1964: 69). His notion developed from a scene on a bell-krater from Syracuse, which depicts Dionysus and a female figure, probably Circe, and satyrs that have been transformed into monkeys. The scene is believed to represent a lost satyr-play. Certain 5th century BCE terracotta monkey figurines from Corinth are remarkably similar to satyr figurines from the same period even engaging in the same activities. Some of the monkeys even ride mules, animals closely associated with satyrs, such as the adorable example in the British Museum. The monkey's legs are splayed as he turns to one side with his arms raised high in the air. Although his face is damaged, a big open-mouthed smile leaves no doubt in the viewer's mind that the little creature is thoroughly elated (**Fig. 103**; Higgins 1986: 116, no. 138). However, the sexual attraction to donkeys seems to be absent in monkeys, even if they are really transformed satyrs. As a general rule, Greek monkeys are portrayed without tails, but two Corinthian examples, one in the British Museum and the other in Copenhagen, both depict monkeys in a wide-legged stance, using their long tails for support. Terracotta satyrs adopt the same stance and are also supported by

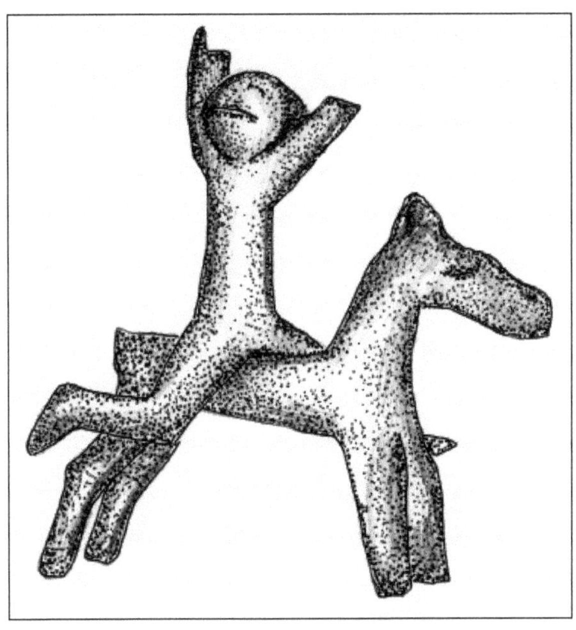

Fig. 103: A terracotta monkey in the British Museum (After Higgins 1986: 116, fig. 138)

[26] I have come across only one example of a masturbating monkey. This is the rather crude Cypro-Archaic terracotta baboon figurine in the Hadjiprodomou Collection, Famagusta, no. 382, possibly found at Komikebir, mentioned above (Karageorghis 1994: 69, fig. 10).

their tails. The satyr in the Goulandris Museum, described above, is a perfect example. The monkeys' bodies also resemble the satyrs in that the heads are disproportionately large and the male genitalia are quite pronounced. The British Museum monkey even employs the common satyr gesture of *aposkopein*. He holds both hands above his eyes, as if trying to protect them from the sun as he gazes into the distance. Hoffman comments that the Copenhagen monkey holds his hands to his face, "as if in amazement at his transformation" (1964: 69).

The notion of a monkey transformation is also applied to the Cercopes in a much later passage from Ovid's *Metamorphoses*. When Aeneas' ship loses its course, he rounds the island of Pithecusae, named for its inhabitants, the Cercopians, whom Zeus had transformed into hairy, speechless, simian creatures (14.90-100, Miller, Trans., 1964). Ovid's play on words in this passage reflects a practice often employed by other Greek and Latin authors. The name of the tribe is derived from the Greek word for monkey, *pithekos*, but the Latin equivalent, *simia*, is implied. These creatures are at the same time *dissimiles* and *similes* to man. Ovid may even have borrowed the idea from Ennius, who wrote of the animal, "simia quam similis turpissima bestia nobis" (*Fragm. Sat.* 69).[27] The Cercopes (tailed ones), as McDermott remarks, had long been a part of the legend of Herakles. In one of the myths, they captured Herakles through their trickery; in others, he captured them. One version of the second myth holds that Herakles killed them, while another relates that their comments so amused him that he let them go. Their name, like that of the monkey, came to be synonymous with deceit, and the legend of their transformation into monkeys seems to have begun in Hellenistic times with Xenagoras (1938: 60). It is also probable that the Romans associated monkeys with Carthage, since a colony of Barbary macaques would have inhabited the area in ancient times. As mentioned above, the island of Pithecussae or Inarime had Punic connections, so Ovid may have had political reasons for reworking the story of the deceitful Cercopes' transformation into monkeys.

Clearly, the Greeks and Romans had enough associations with monkeys to create caricatures of them in their art and literature, but how much they knew of the actual animals has yet to be analyzed. As previously mentioned, in Greek fables, monkeys are referred to as *pithekoi*, while in Latin, the general term is *simia* or *simius*. Other terms were used in literature, however, and Aristotle was the first to attempt a classification of the three species of monkeys known to him in a passage from *Historia Animalum*, 2.8-9. These were the *pithekoi*, the *keboi* and the *kynokephaloi*. McDermott translates these as the ape, the monkey and the baboon. Of course, Aristotle would not have been aware of apes, and his description of the three species should not be confused with modern terminology. He maintains that these three animals share the traits of humans and quadrupeds and explain that the *kebos* is a *pithekos* with a tail. The *kynokephalos* is the same shape as the *pithekos*, but it is bigger, stronger and has a dog-like face. Clearly, Aristotle uses the term *kebos* to refer to any long-tailed monkey, probably either vervets or patas monkeys imported from Egypt. *Kynokephalos* is the general term for baboons, as it is today, and *pithekos* seems to refer to the tailless Barbary macaque as well as monkeys in general. The majority of ancient authors, however, were not as careful with their terminology.

Pliny, too, discusses monkeys in a passage from his *Natural History*, 11.246, which mainly echoes the description provided by Aristotle in the passage cited above. It is worth repeating here, since it reflects a Greco-Roman observation of the monkey's biological similarity to humans that persisted well into the modern era (Personal translation):

> nam simiarum genera perfectam hominis imitationem continent facie, naribus, auribu, palpebris, quas solae quadripedum et in inferiore habent gena, iam mammas in pectore et bracchia et crura in contrarium similiter flexa, in manibus ungues, digitos longioremque medium. Pedibus paulum differunt; sunt enim ut manus praelongae, sed vestigium palmae simile faciunt. Pollux quoque iis et articuli ut homini; ac praeter genitale, et hoc in maribus tantum, viscera etiam interiora omnia ad exemplar.
>
> Indeed the family of monkeys has a perfect imitation of a person in their face, nostrils, ears and eyelashes. For they are the only quadrupeds with eyelashes, and they even have them on the lowerlids. They also have breasts on the chest, and similarly their arms and legs bend in oppposite directions. There are nails on their hands and a long middle finger. They differ a little in their feet, for these are elongated like their hands, but make a palm-like foot-print. They also have a thumb and knuckles like a person; and besides genitals, and this is in the males only, they also have all the same pattern of internal organs.

Galen, writing in the 2nd century CE, took the practice of observation further by dissecting and vivisecting a number of animals, including monkeys.[28] McDermott asserts that dissection of human corpses was rarely permitted and that Galen believed the internal structure of humans was the same as monkeys. This, of course, led to many mistaken assumptions about the human body, but Galen's teachings were so widely accepted that his authority was not contested until the 16th century CE (1938: 93). In *De Anatomicis Administrationibus*, 9.13, Galen discusses the selection of monkeys for dissecting the brain (Duckworth, Trans. 1962):

> The dissection is best made in apes, and among apes in such a one as has a face rounded to the greatest extent possible amongst apes. For the apes with rounded faces are most like human beings.

McDermott comments that the monkey that Galen most commonly used was the Barbary macaque (1938: 94), which explains the common mistranslation of ape. The remainder of the passage lends support to the identification. After several long sections of graphic detail on the correct method of removing the skull, Galen suggests that students practice the same procedure on a living animal. He emphasizes that pigs or goats are the best choice and

[27] How similar to us is the simian, the most base of animals (personal translation).

[28] McDermott maintains that Rufus of Ephesus dissected monkeys before Galen (1938: 93).

that monkeys should be avoided for two reasons. First, the monkey's expression is unsettling. Second, it is best to have an animal that cries out with a loud voice, which monkeys will not do (*AA* 9.18-19). Unfortunately, this, in my mind, confirms that Galen did vivisect macaques. These monkeys are known to put on a brave front and generally do not cry out when they are engaged in a fight, even if they have been seriously injured. If the animal were restrained and mutilated, it would most likely grimace and, at the most, make a low kecking sound. It is interesting, however, that Galen found the grimace disturbing. One might surmise that he felt some empathy for this human-looking creature.

McDermott summarizes some of the more noteworthy of Galen's findings and provides a list of every reference made to monkeys in his writings, so I will not linger on that subject here.[29] It is worth mentioning, however, that Galen was aware of different species of monkeys, which he named as: πίθηκος, λύγξ, σάτυρος, κυνοκέφαλος and κηβος. The *pithekos*, *kynokephalos* and *kebos* have already been identified—although some ancient authors have more to say about the *kebos*—but the *lynx* and *satyros* are problematic. As McDermott remarks, the lynx is usually cat, but Pliny mentions it along with *sphinx*, as a type of monkey from Ethiopia (1938: 95; *HN* 8.30, Personal translation):

> Lyncas vulgo frequentes et sphingas fusco pilo, mammis in pectore geminis, Aethiopia generat, multaque alia monstris similia...
>
> Ethiopia produces many lynxes and sphinxes with dark hair and breasts on the chest, and many other such monsters...

The sphinx, with dark hair and breasts, could well be the gelada, but this is pure speculation. Pliny goes on to talk of winged horses and, therefore, probably received much of his information from travellers' tales rather than observation, especially since Aelian identifies it as an Indian monkey (*NA* 16. 15). An identification of the species meant by 'lynx' cannot be suggested from the passage, and Galen's descriptions are of little help because he seems only to mention their similarities to other monkeys without naming distinctive characteristics (McDermott 1938: 96). The *satryros* is mentioned as an Indian species by both Pliny and Aelian; the latter, citing Cleitarchas, presents a strange portrait of Indian monkeys (*pithekoi*) in general. He relates the story that when Alexander and his army arrived in India, they were terrified by the sight of what they believed to be a great army waiting to ambush them. The army, however, turned out to be a large troop of monkeys standing upright. Aelian comments further that these monkeys cannot be captured by nets or dogs but only by their own imitative nature. When the monkeys see a hunter putting kohl under his eyes, they try to imitate the action, but the hunter tricks the animals by adding bird-lime to the kohl, which glues their eyes shut. Monkeys are also tricked into putting on lead shoes or gazing into trick mirrors, which also glue their eyes shut (*NA* 17.25). As Jennison comments, the account is probably fictitious (1937: 148).

[29] McDermott lists every reference to monkeys in Galen's works, as cited in Kuehn's volume (1938: 94, note 25).

The behaviour of Egyptian baboons seems to have been of special interest to Aelian, who seems to have a rather favourable view of them (*NA* 10.30, Schofield, Trans. 1959):

> If a baboon finds some edible object with a shell on it...it strips the shell off and cleans it out, after first breaking it most intelligently, and it knows that the contents are good to eat but that the outside is to be thrown away. And it will drink wine, and if boiled or cooked meat is served to it, it will eat its fill; and it likes well-seasoned food, but food boiled without any care it dislikes. If it wears clothes, it is careful of them...If you put it while tiny to a woman's breast, it will suck the milk like a baby.

Strabo also commented on the Egyptian baboon, saying that it was worshipped at Hermopolis but that the *kebos* was worshipped near Memphis (17.812). McDermott maintains that the use of the word *kebos* was either an error or intended as a synonym of *kynokephalos* (1938: 37). However, as discussed in Chapter 1, the Egyptians mummified smaller monkeys as well as baboons, and vervets acquired an association with Bes. Therefore, Strabo may well have observed long-tailed monkeys used in a religious context.

One traveller's tale that is of extreme interest comes from the *Periplus* of the Carthaginian explorer, Hanno. Because Hanno was a common name throughout Carthaginian history, the author of the text is difficult to identify with any certainty. The only sure knowledge we have of this Hanno is that he was a king, who lived before Herodotus. The voyage seems to have taken place in the 5th century BCE, as the result of concerns that the Greek colonial expansion would interfere with Carthaginian trade routes. Considerable controversy exists over the extent of Hanno's voyage, but it is doubtful that he circumnavigated the whole of Africa (Oikonomides 1977: 9-16). The last passage of text refers to strange creatures that inhabit an island in a bay known as the Horn of the South. The natives who serve as guides refer to the beings as Γορίλλας. Oikonomides, using the Codex Palatinus Graecus 398, translates the episode in this way (1977: 29, 56r.87-100):

> The biggest number of them / were females, with hairy bodies, which our [Lixitae] / interpreters called "Gorillas." Chasing them, we could / not catch any of the males, because all of them escaped // by being able to climb steep cliffs and defending them- / selves with whatever was available; but we caught three females / who bit and scratched their captors and they did not want / to follow them. So we had to kill them and flayed them / and we brought their skins to Carthage.

The translator comments that scholars have been careful to avoid identifying the gorillas in Hanno's voyage as modern-day gorillas, primarily because the animals do not have a habitat in the area of Sierra Leone, where the incident is generally believed to have taken place (Oikonomides 1977: 35). Indeed Janson argues that Hanno made no mention of the animals' size and that the gorillas could have been any primate. Baboons, as he rightly comments, commonly live near rocky cliffs. He also argues that even if Hanno did encounter great apes (chimpanzees, rather than gorillas), the skins he brought back to Carthage probably contributed very little to the ancient knowledge of anthropid apes (1952: 328-29). The account, however, did contribute something to Greco-Roman legends of fantastical, monkey-like beings. Pomponius Mela, for example, writing in the time of Claudius, cited the island of hairy, wild women that was visited by Hanno (McDermott 1938: 52).

In daily life, Greeks and Romans were also familiar with monkeys as pets. Most ancient authors, however, are not impressed with the animals and ridicule their owners. In a letter to Atticus, Cicero writes from Laodicea that he encountered a certain idiot, P. Vedas, who was travelling with a large entourage of servants, onagers and a pet baboon (*Ad Atticum* 6.1.25). As McDermott remarks, Cicero is polite to Vedas only because of his connections to Pompey, and his contempt of his pets is equal to his contempt of the man (1936: 155). Elsewhere, Martial suggests Cronius is as ugly as his monkey (7.87.4, personal translation):

> si Cronius simile cercopithecon amat…
>
> If Cronius loves a monkey as [ugly] as himself…

Despite these negative portrayals of monkey keepers, it seems to have been acceptable for children to love their pet monkeys. McDermott catalogued three funeral stele that depict monkeys. The stele of C. Julius Saecularis depicts a boy surrounded by his beloved pets, which include a dog and monkey that clutches the boy's garments (1938: 302, no. 507). He also notes that Pindar, *Pythean* 2.72-3, said that monkeys were always beautiful to children. His comment is directed against a flatterer at the court of Hiero. The immature judgments of children are contrasted with the sound judgment of the mature man, who recognizes a sycophant for what he is (1938: 132).

Primates were also, on at least one occasion, brought into the arena. Pliny mentioned that one of the animals Pompey imported for the games of 55 BCE was the Ethiopian *cephi*, which was not seen in Rome again (*NH* 8.28). McDermott remarks that although Pliny did not say the animals were used in combat, they probably were used in a *venatio* rather than an exhibition (1938: 71). While this seems probable enough, his suggestion that Pliny's *cephi* was actually a gorilla is highly unlikely. If an unusual animal were introduced to Rome from Ethiopia, this could well have been the gelada. These baboon-like animals are fairly strong, and an adult male, with its large canines, could probably put on a good fight if provoked.

Additionally, monkeys were used in Rome as part of the punishment for parricides. The individual was forced to wear wooden shoes and wolf-skin cap. He was then beaten with rods and sewn into a bag with a viper, a monkey, a dog and a rooster. The sack was then thrown into the sea or river, where the offender and quarrelling animals would drown. McDermott explains that the ritual was primarily intended to remove an evil portent from the city and that the animals were selected for their evil character (1938: 153-4). Apart from embodying ugliness and conceit, monkeys were considered bad omens. Suetonius claimed that Nero, shortly before his downfall, dreamt the hindquarters of his favourite horse turned into a monkey's rump (*Nero* 46.1).

In Roman times, Egypt and the regions to the south were still considered mysterious lands, full of exotic animals, with monkeys featuring as prominent members of the local fauna. One of the most famous and beautiful depictions of an Egyptian landscape is the Nile mosaic from Praeneste, discovered in the 17th century CE. The dating of the mosaic is problematic, but it belongs to a genre that became popular in the 1st century CE and was originally connected to the ruins of an architectural complex upon which the Temple of Fortuna Primigenia was built. The complex, however, probably dates from 125-100 BCE, and Whitehouse accepts this as a likely date for the mosaic as well (2001: 71-2). From the time of its discovery, the mosaic has been severely damaged, restored and moved. However, a series of 17th century watercolours provide valuable evidence regarding the original appearance of the scene. As Whitehouse comments, the mosaic provides a bird's eye view of the Nile Valley, with the southern regions at the top and the northern Delta below (2001: 71). At the top, Nubian hunters, standing on a rocky hill shoot arrows at their prey. Below them are a variety of exotic animals, most of which have their names beside them in Greek. The lower half of the scene depicts the towns of the Delta and the yearly festival of the inundation. The scene contains four monkeys whose identification poses problems. McDermott remarks the labels have caused confusion because the artist did not have precise understanding of the animals he was portraying (1938: 284). In some cases, the monkeys may even have been fictional. One section of the scene depicts that ever elusive primate known as the *satyros*. The inscription next to the animal has been incorrectly restored as *CATTYOC*. Indeed, it is a little difficult to positively identify the creature as a monkey, but it could hardly be anything else. Its fingers are well defined, and its large feet have long, monkey-like toes. The animal sits with one leg drawn up to its chest, while the other is stretched out in a relaxed manner. The tail is long and has a tufted end. Its face is rather cat-like, but clearly some type of simian is intended .

Another section depicts a greyish-black long-tailed monkey sitting on a rock in an extremely naturalistic manner. It looks downward at giant crabs rising from the water below the rocks. The animal has a very shaggy appearance overall, and, once again, may not depict an actual species. To the right of this monkey is a tree with two animals in it. On an upper branch, a large, apparently tailless, monkey is curled up with its head and hands resting on its knees. A long-tailed animal, with the label CΦINΓIA, walks along a branch to the lower left. This animal could be either a monkey or a cat. It has a lion's tuft at the end of its tail, and a lion-like mane. The front paws do have rather long fingers, but the back feet are too short to belong to a monkey. In the upper right-hand corner of the mosaic, a readily identifiable monkey without a label, sits in a typical crouching position. It seems to be smiling as it looks over its shoulder at a fantastical four-footed creature with the body of a lion, the hooves of a goat or cow and the face of a human. This creature is identified as a *MONOKENTAYPA*. The portrayal of the monkeys and other creatures in the Nile mosaic, then, suggest that Romans had only a partial familiarity with the animals. They were readily familiar only with a small number of species but knew that others existed. The artists let their imagination fill in the missing information. Therefore, some monkeys mentioned by ancient authors, such as the lynx, sphinx and satyros, will probably never be identified by modern scholars because the animals themselves were only half real. Their rarity put them in a mythical realm. Hazy memories of a species once glimpsed in the arena or on a trip to Egypt produced the legends and peculiar iconography that will probably always leave modern readers guessing.

Since the Egyptian/Ethiopian source of monkeys was an object of fascination for the Greeks and Romans, a brief exploration of their understanding of the simian's role in Egyptian religion is in order. Once again, Greco-Roman ideas about the significance of baboons and other monkeys reveal that much was lost in translation. The most detailed account comes from a 4th-5th century CE text, known as the *Hieroglyphica*, attributed to Horapollo Niliacus. The text was probably originally written in Coptic and later translated into Greek. Horapollo had probably never actually learned hieroglyphs and early attempts at deciphering the language based on his comments were usually futile. Nevertheless, the first book of the Hieroglyphica contains three sections describing the use of the *kynokephalos* in hieroglyphs, which, rather surprisingly, relate a number of their roles accurately. There are, of course, a few glaring instances of miscomprehension. As Grafton comments, though, the work as a whole sheds light on the intellectual life of antiquity, at a time when attempts were made to combine Eastern wisdom with Greek philosophy (1993: xv). Therefore, it seems worthwhile to discuss several of Horapollo's claims listed in the three passages on baboons for a better understanding of the Greco-Roman view of Egyptian beliefs about the animal.[30] The passages correctly identify the baboon's connection with the moon but contain a peculiar mixture of true and nonsensical assertions:

1. In hieroglyphs, baboons represent the inhabited earth, a priest, anger, or a diver. They represent anger because they are the most irritable of animals and the diver because they do not appear dirty after a swim.
2. When the moon is darkened, baboons do not eat and bow their heads as if in mourning. Females bleed from their genitals.
3. Baboons are kept in temples because their behaviour indicates the conjuction of the sun and moon.
4. Baboons do not die in one day. Part of them dies everyday for seventy-two days until the process is complete. Each day a funerary rite is held in honour of the dead part.
5. Some baboons can write, and priests test them by giving them a tablet and pen. They are sacred to Hermes, god of writing.
6. Priests do not eat fish because of the baboon.
7. Baboons are born circumcised.
8. Baboons represent the rising moon when they are shown with raised hands and wearing a crown.
9. Seated baboons represent the equinoxes. At this time of the year, baboons urinate twelve times per day and per night. This is why water clocks are made in the form of baboons. They also made a loud racket every hour.

As discussed in Chapter 1, Thoth in his baboon manifestation had strong connections with the moon. This association was as strong as ever even in the late Roman period, and the Greco-Roman philosophers (if not the Egyptians themselves) sought an explanation in the baboon's behaviour. Female baboons do, indeed, have menstrual cycles as frequently as human women, and this approximately monthly cycle could easily be interpreted as part of a natural rhythm based on the phases of the moon. Additionally, baboons do tend to find a resting place before sunset and fall asleep as soon as night falls. In the wild, this is usually a high place, such as a cliff, where they will be safe from predators. They usually sleep in a seated position with their heads bowed, which could be mistaken for the human gesture of mourning.

The strange belief about the baboons' slow death demonstrates a partial understanding of Egyptian beliefs and practices. McDermott cites Iamblichus' *De mysteriis* 8.3 as evidence that the Egyptians divided the heavens into seventy-two parts, and the idea that baboons in temples die over a period of seventy-two days refers to the practice of mummification. The Rhind papyrus, dating from the Augustan period, specifies seventy days for the preparation of the body and another two days for the funeral (McDermott 1938: 44-5). Although the papyrus does not refer to baboons, it seems these creatures were accorded the same privileges granted to humans, especially those baboons that were considered the living incarnation of Thoth. The association of baboons with writing also had strong roots in Egyptian religion, as discussed in Chapter 1, since Thoth was the god of scribes, and his assimilation with Hermes is evident here. The idea that baboons could write is also found in Aelian, *NA* 6.10, in which he comments that they were taught to write under the Ptolemies. Although the animals could not have actually been taught to write, they would probably have enjoyed playing with the ink. It is not inconceivable that they may have been given writing implements and encouraged to use them in their own fashion.

Their connection to priests through the belief that they did not eat fish is understandable, since the wild baboon's diet consists mainly of fruit, insects, eggs and occasionally meat from animals they catch. In Egypt, they would not have been given fish because it was considered an unclean food associated with Seth/Typhon, as mentioned by Plutarch in *Moralia* 353d-f and 729a. Additionally, as McDermott comments, Egyptian priests would still have been circumcised in the Greco-Roman period, even if ordinary citizens were not (1938: 45). A baboon's erect penis would have a circumcised appearance because it does not have a foreskin covering the glans. When not in a state of arousal, the member simply retracts into the folds of flesh above the testicles.

The use of the baboon as a determinative for the word 'to be furious' was also mentioned in the first chapter, so Horapollo seems to have interpreted that perfectly. His suggestion about the swimming baboon, however, is not easily explained, but the animals are not afraid of water and do have sleek coats that dry quickly. Boas comments that Horapollo's assertion that the baboon with upraised arms represents the moon was actually useful in deciphering an inscription (1993: 54). Finally, the use of baboon-shaped clepsydria must certainly have led to the rather bizarre statements about the baboon's behaviour during the equinoxes. Horapollo seems not to have understood

[30] See Boas' translation 1993: 14, for the complete three passages.

that the baboon's connection to time is the result of its association with Thoth, as the calculator of time and the seasons.

Clearly, in Roman Egypt, the worship of Thoth was changing. As discussed in Chapter 1, by the Ptolemaic period, the cult centres were requiring more and more animals to be sacrificed, which may have led to a depletion of hamadryas baboons. Barbary macaques were imported from the west as substitutes, and the mummified bodies of the animals reveal that they suffered from poor health, probably as the result of a vitamin D deficiency. Images of the god in his baboon manifestation continued to be made in the form of statues and water-clocks, as evidenced by Horapollo's text, and Thoth continued to be portrayed as a baboon in vignettes from the Book of the Dead. Thoth had also begun to merge with the Greek god, Hermes, and acquired the epithet Trismegistis, the 'thrice great.' This syncretism with Hermes grew stronger in Roman times and led to the development of Hermetic cults, whose beliefs were based on the fusion of religious and philosophical traditions of the two cultures. New personalities and new associations with other deities were created as a result. Hermes' son, Asclepius, became identified with the Egyptian Imhotep, the deified architect of Djoser's pyramid. Isis and Horus, who had long been associated with Thoth, retained their relationship, but an intermediary between the two, *Kore kosmou* or Kamephis, was also introduced. Ammon, based on the Egyptian Amun, came to be considered an early king, and Thoth acquired another son, Tat, based on a misspelling of his name in Greek. Other personalities included the winged serpent, Agathos Daimon; Poimandres, the divine intellect; and Bitys, the priest who translated the sacred writings for Ammon (Fowden 1993: 32-3). By the Roman period, Thoth was one of the most popular of Egyptian gods and his powers were numerous. Most of his functions ultimately derived from his traditional role as a lunar deity, who regulated the seasons, as well as his role as the scribe of Re. Fowden maintains that Thoth "presided over almost every aspect of the temple cults, law and the civil year, and in particular over the sacred rituals, texts and formulae, and the magic arts that were so closely related to him" (1993: 22). As a deity associated with esoteric wisdom, magic, and medicine, who acted as psychopompos in the Duat, Thoth's assimilation with Hermes, who filled similar roles in Greek religion, was completely understandable. Through his many conflicts with Seth and Babi, Thoth also had a reputation for revealing the truth by means of trickery, which further connected him to Hermes (1993: 24).

The corpus of Hermetic discourses, produced mainly in the Greek city of Alexandria, were said to have been written by the hand of the Thoth. An exploration of the ideas expressed in the writings is not suitable here, but aspects of the cults are worthy of mention. The teachings take the form of conversations between Poimandres (*Nous*), as teacher, and Hermes Trismegistus, as student, or Hermes, as teacher, and Asclepius or Tat, as student. The form is reminiscent of Platonic dialogues, and the philosophy has been compared with late Platonism and Gnosticism. The teachings contained a message of salvation through knowledge, which was restricted to the elite (Fowden 1993: 188-9). The influence of the writings, however, was not limited to Alexandria or to Greek citizens. Other Egyptian texts from the Roman period attributed to Thoth were written in a similar form but retained an entirely Egyptian content. It seems the mysteries that were once restricted to the temples came to be transmitted in a more personal way, as if from father to son (Roberts 2000: 198-9).

In art, Thoth's baboon iconography began to incorporate the personalities of the Hermetic tradition. A 1st century CE stele in the Allard Pierson museum depicts Thoth as a baboon wearing a disc and lunar crescent; he sits on an altar holding a plume in one hand. His presence is reinforced both by a seated ibis floating before his raised right hand and by the depiction of the god in his anthropomorphic ibis-headed form, also facing the baboon. The infant Horus approaches the baboon from the opposite side of the scene. Above them, Agathos Daimon, depicted as a double-headed snake, spreads his wings (Salaman *et al.* 1999: pl. 3). Another 1st century CE sculpture from the Allard Pierson museum depicts Thoth-Baboon seated on an altar, wearing a disk with the image of the snake, Agathos Daimon. Thoth holds a scroll in his hands, and his presence is reinforced by the walking ibis carved in relief on the altar (1999: pl. 4; **Fig. 104**). Other images of Thoth that were not necessarily connected to the Hermetic cults also continued to be made. Lamps in the form of Thoth-Baboon, for example, were a common item in the 2nd-3rd century CE (Fjeldhagen 1995: 179, no. 177).

From this account of the monkey's reception throughout the various regions and time periods that comprise the Greco-Roman world, it is evident that the little creature

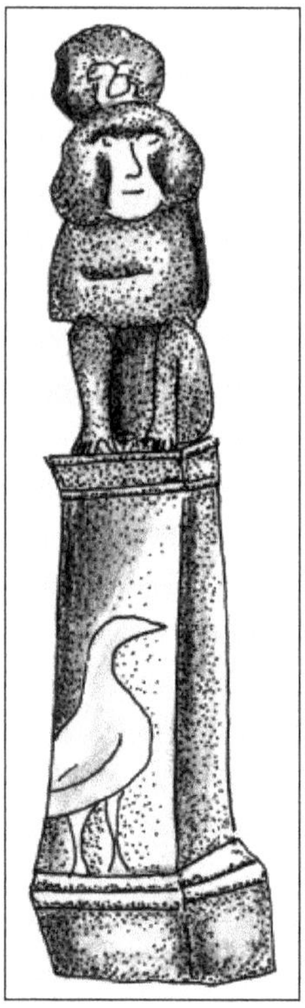

Fig. 104: Thoth as a baboon, wearing a disc with Agathos Daimon in the form of a snake (After Salaman 1999: pl. 4)

faced considerable hostility. Initially, though, a few Greek sites, influenced by Egypto-Phoenician traditions, greeted the monkey with a certain amount of cordiality. The animal's apotropaic functions and value as an exotic rendered its image desirable. Pithemorphic aryballoi from eastern sites depict stylized but attractive animals that would in no way inspire ridicule, and even the stylized Boeotian figurines may have had value as votive offerings. Nevertheless, early references to Aesopic fables depicted the monkey in a negative manner, and by the 5th century, the monkey had completely lost its respectability in Greco-Roman culture. Perhaps the ready availability of monkeys from North Africa and their association with the area of Carthage led the Romans, in particular, to hold the creature in contempt, but it is really the Greco-Roman notion that beauty is defined by the human form that led to the monkey's downfall. It appeared beautiful only to immature children and foolish, vain adults. It was a reflection of the worst human qualities and not a creature to be emulated, as the superstitious Egyptians believed.

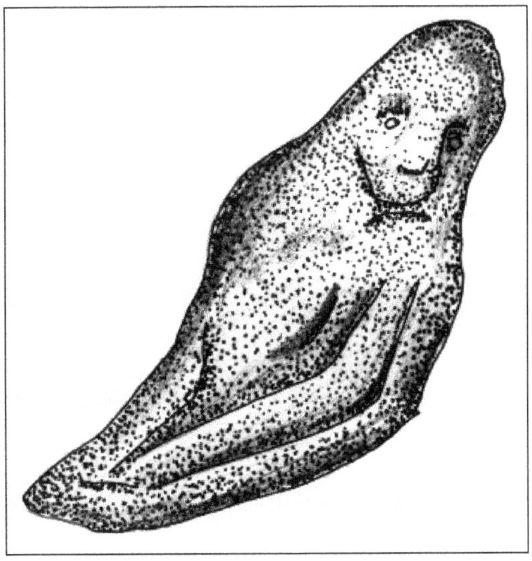

Fig. 105: A tragic monkey
(After Kozloff 1981: 144, no. 122)

It seems appropriate that a 5th century eastern Greek coroplast, an artist who would not have been highly esteemed in his own time, managed to capture so eloquently the monkey's tragedy. A rough mould-made terracotta, once again from the Mildenberg collection, depicts a little monkey sitting with his legs outstretched and his hands resting stiffly on his knees. His head tilts down, and his eyes and muzzle, despite the crude rendering of his body, are well formed. His overall expression is that of grief. Mitten remarks that its *ethos* is "as effectively moving as any of the moods visible in the protagonists in the metopes of the Temple of Zeus at Olympia or contemporary Attic red-figured vases" (**Fig. 105**; In Kozloff 1981: 144, no. 122). It is impossible to know why the coroplast created such a sad monkey—perhaps it is a caricature of tragic figures—but the creature has every reason to mourn its fate. The Greeks and Romans left a legacy of speciesism that would tarnish the monkey's image well into the 20th century CE. As demonstrated in the following chapter, our sense of superiority over monkeys, those 'basest of animals most similar to us,' had its roots in ancient times and has led to the near extinction of a number of primate species. Monkeys and apes, victims of the Greco-Roman tradition, would be cursed by Christians, hunted by explorers, tortured in labs and exploited in films.

Chapter 5
THE GRECO-ROMAN LEGACY

This chapter goes beyond the ancient Mediterranean to explore the ways in which Greco-Roman perceptions of non-human primates influenced Western culture from early Christian times to the present. To sum up the set of associations, monkeys were portrayed in literature as foolish, vain, ugly and deceitful. In dreams, monkeys were considered an evil omen. As pets, they brought scorn upon their owners. Their ability to entertain audiences through mimicry won them a combination of praise and ridicule. Monkeys made people laugh, but only because they were considered poor imitations of human beings. Naturally, this led early Christians to view monkeys with considerable suspicion, and the hideous portrait they created would harm the animal's reputation well into the 21st century. Early philosophers and doctors also recognized the value of the monkey's anatomical similarity to humans and used them in experiments involving dissection and vivisection. Galen's view of human anatomy, based on an exaggerated similarity with that of the Barbary macaque, was widely accepted throughout the Middle Ages and was not, in fact, called into question until the 16th century. Sadly, the use of monkeys in scientific experimentation, including vivisection, continues into our own day. This chapter begins in the early Middle Ages and is greatly indebted to the admirable work of H.W. Janson (1952), *Apes and Ape Lore in the Middle Ages and Renaissance*, which explores the roles of monkeys in art, culture and literature up to the discovery of anthropoid apes. The study continues with a brief glimpse at the manner in which simian iconography was used to create caricatures of ethnic minorities, especially after Darwin published his famous treatise, *Origin of the Species*. In popular culture, monkeys and apes have also played a variety of roles in film and literature, from giant lecherous monsters to cuddly quasi-humans.

Perhaps Ennius' description of the monkey as *simia quam similis turpissima bestia nobis*[31] (Cicero, *De natura deorum* 1.35.97), best describes the Greco-Roman attitude toward monkeys that persisted into the Christian period. Because monkeys were considered base caricatures of the human form and the term *simia* had acquired all the negative associations mentioned above, early Christians applied the word to all enemies of Christ. Janson maintains that the most damaging account of monkeys comes from the *Physiologus*, a Medieval compendium of Christian zoology that seems to have drawn much of its information about monkeys from Horapollo's description of the functions of the sacred baboon (Janson 1952: 16-17). The monkey and the onager are named as the two animals that announce the equinox. The onager, like the baboon in the *Hieroglyphica*, does this by braying twelve times, while the monkey urinates seven. The night, symbolizing pagans, becomes equal to the day, symbolizing Christians, and the devil rejoices. The monkey's tailless form, an object of ridicule in Greco-Roman literature, then led the author to liken it to Satan (Janson, Trans. 1952: 16-17):

> Habuit enim initium, finem autem non habet (hoc est caudem); in pricipio autem fuit unus ex archangelis, finis autem eius nec invenitur. Beneque simius, non habens caudem, sine specie enim est; et turpe in simio, non habentem caudem; sicut et diabulus, non habet finem bonum.
>
> He had a beginning, but he has no end (that is no tail); at the outset he was one of the archangels, but his end is not in view. Now the ape, not having a tail, is without species, and his rear, without a tail, is vile; like the devil, he does not have a good end.

The animal described here is the Barbary macaque, which also posed problems for Aristotle. His inability to definitively categorize the animal is the obvious source of the *Physiologus* author's assertion that the simia is without species. The *pithekos* is a quadruped that shares qualities with humans. Its face is similar to a human's, but it lacks both the buttocks of a human and the tail of a quadruped (*HA* 2.8). Furthermore, the medieval author would have retained the Greco-Roman disdain for the Egyptian practice of depicting gods in animal form. Janson remarks that after Bishop Theophilus had led the destruction of pagan temples and idols during the riots at Alexandria in 391 CE, he ordered the preservation of one cult statue of a baboon, which was to be publicly displayed as a reminder of the depravity of pagan practices (1952: 17).

[31] How similar to us is that basest of animals, the simian (personal translation).

In addition to its lack of tail, Janson maintains that the monkey's natural ability as an imitator led to further comparisons with Satan. He cites a Latin translation of Isaiah 14.14, in which Lucifer claims he will be similar to God, *similis ero altissimo*. The devil became *simia dei* and the monkey became *figura diaboli* (1952: 20). Early Christian works reflect this new association. The 9th century Stuttgart Psalter, which illustrates Psalm 77, depicts a deformed monkey kneeling above the devil, who is portrayed as a rather satyr-like being with horns and hooves (1952: pl.1). The marvellous 11th century tympanum of Puerta de las Platerias at Santiago de Compostela even depicts one of the demons from the Temptation of Christ as a winged monkey (1952: pl.2). Finally, although the monkey was later demoted to the status of a mere sinner, Janson remarks that it retained its association with the devil as a 'recessive character' into the 17th century. He cites a painting, attributed to Domenico Feti in evidence. An innocent looking monkey, representing the devil, sits behind St. Dominic, holding a candle. The little creature has burned its hand and sucks its fingers with a sad look. The painting relates the story of Satan's attempt to lead Dominic astray, but the clever saint tricks the devil and temporarily makes him his servant (1952: 22, pl. 3). It seems possible that the role of the monkey as the dupe of more clever animals in Aesopic fables may also have played a role in this story. La Fontaine, writing in the 17th century, reworked a number of the fables related in Chapter 4, including the monkey and the dolphin. The monkey, who attempts to trick the dolphin by claiming to be a man, is caught lying and drowned (4.7). Although Janson does not make the connection with this story, it is easy to see how folktales in which the hero tricks the devil could be conflated with the monkey in Aesopic fables.

By the 12th century, the monkey began to lose some of its frightening power as the *figura diaboli*. Instead, it was to become the symbol of human degeneration, a step up from the Face of the Devil. Janson suggests that the change may have occurred because monkeys were becoming more common in Europe, and people had difficulty equating comical performing monkeys and their trainers with anything as dreadful as Satan (1952: 30). He remarks also that Classical literature experienced a revival at this time. The Aesopic fable of the monkey and her twins gained considerable popularity as an allegory of the sinner and replaced the *Physiologus'* portrait of the devil-monkey. Since Medieval authorities accepted Avianus' version of the fable, the idea that a monkey holds its favourite child to its breast and its hated child on its back became an accepted account of the monkey's behaviour. In time, this factoid acquired a moralistic interpretation. The beloved son that is dropped came to symbolize bodily pleasure, while the son that clings to her back symbolized the goods of the soul. Later, the hated son came symbolize the sins that cling to one's back and cannot be shaken (Janson 1952: 31-3).

As the symbol of the sinner, the monkey was also placed both in scenes of Adam naming the animals and in scenes of the Fall. Janson remarks a certain similarity between these scenes and late antique depictions of Orpheus charming the animals with his music, in which the monkey is inserted playing a musical instrument. He argues that monkey is more than just comical element—it is the only animal that resists the hero's spell (1952: 108). Representations of the monkey eating an apple reflect this rebellious aspect of the sinner who does not bend to the will of Adam and, indirectly, to the will of God. Also at this time, the monkey is equated with the female sinner, and it is referred to by the feminine *simia* rather than *simius*, which may have led to its association with Eve and her role in the Fall (Janson 1952: 109). A relief by Ludwig Krug, dating from 1514, makes the connection clear. Adam and Eve stand under the tree, contemplating the apple in Eve's hand. A large monkey walks between them munching on an apple that has fallen from the same tree. The snake that coils around the trunk is so small in comparison that it is hardly noticeable (1952: pl. 18).

Later, the monkey came to be used in various non-religious, although still negative, contexts. The monkey could symbolize foolishness and vanity, just as it did in the Greco-Roman tradition. Janson explains that 15th century depictions of monkeys gazing into mirrors reflect the concept of *vanitas* and comments that the image is also reminiscent of Aelian's assertion, cited in the previous chapter, that Indian monkeys were caught by gazing into trick mirrors (Janson 1952: 212). The iconography also came to symbolize the monkey's foolishness at flying into a rage when it realized its reflection was not another monkey (1952: 213). A Neoplatonic view of monkeys has also been read into the roughly carved monkey behind the legs of Michaelangelo's Dying Slave. The monkey is believed to represent the 'Lower Soul,' which is enslaved by matter (1952: 297).

Perhaps the most fascinating legacy is the Greco-Roman conception of human anatomy based on the dissection of monkeys. Galen's ideas concerning human anatomy remained unchallenged until the 16th century, when Vesalius caused considerable controversy by rejecting a number of the ancient author's assertions in his 1543 work, *De humani corporis fabrica librii vii* (Janson 1952: 358). A caricature of the Laocoon group, attributed to Boldrini of Titian's workshop, in which Laocoon and sons are all depicted as monkeys, may satirize the refusal of Vesalius' opponents to accept that Galen made some serious mistakes. The three monkeys may depict the ideal human form according to Galen's description of human anatomy (Janson 1952: 355-364, pl. 56).

More recently, after Darwin published his theory of evolution in 1859, a furore arose over the suggestion that human beings evolved from a species of ape that has since become extinct. On Nov. 25, 1864, Benjamin Disraeli responded to the new theory with the now famous words, "The question is, is man an ape or an angel? Now I am on the side of the angels."[32] In the wake of the controversy, a number of ethnic groups, including the Irish, were caricatured as apes, while the English and other Germanic races were depicted as the descendants of angels. The practice, however, was not new. Earlier proponents of the 'science' of physiognomy did much to emphasize the difference in facial features between northern Europeans and all the other races of the world, which were considered inferior. Pieter Camper, believing that a study of facial angles would effectively demonstrate

[32] Punch, or the London Charivari (Dec. 10, 1864), caricatured Disraeli as a rather feminine angel (Curtis 1971: 106, fig. 46).

a hierarchy of intelligence, created a chart to demonstrate the intervals of development among monkeys, apes and various human races. The chart depicts the African as closer in appearance and, by implication, intelligence to an orang-utan than to a European. The man's features have been exaggeratedly simianized (Curtis 1971: 7-8, fig. 1). Camper's measurements had a considerable effect on the caricature of the Irish from the 1840s onward. The Punch cartoonist, J. Kenny Meadows, may have had such an image in mind when he created the caricature of Daniel O'Connell as a simianized Frankenstein (Curtis 1971: 32, fig.6).

The earlier discovery of anthropoid apes no doubt contributed to subconscious fears concerning a 'primitive' state to which we might revert and monstrous beings that resemble us. Our modern-day gorilla was, of course, named after the aggressive, hairy beings described by Hanno, which could not have contributed positively to its reception in the west. Perhaps the best known fictional gorilla is the giant King Kong, whose image is frightening and sympathetic at the same time. His primitive condition is enhanced by the mysterious land in which he is discovered. He lives in a previously unexplored jungle, inhabited by dinosaurs and other terrifying creatures that should have been extinct for millions of years. This giant ape, like baboons in Greco-Roman tradition, is drawn toward a human female, and sexual tension is omnipresent in the film. Scenes of Kong stripping off Fay Wray's dress and then sniffing his fingers were originally cut, but the sexual symbolism in the gorilla's climb, with lady-love in hand, "to the tallest erection in New York—a pinnacle of rape…is as old as the first menhir set up in Neolithic times" (Annan 1975: 25).

In addition to this perceived sexual aggression, apes were thought of as indifferent to human life or even murderous. When Kong pulls the wrong woman out of a window, he callously lets her fall to her death.[33] In the 19th century tale by Edgar Allan Poe, *Murders in the Rue Morgue*, an orang-utan steals its owner's razor and commits two brutal murders. Such portrayals of great apes are reminiscent of the depiction of the giant simian monster on the Cypro-Phoenician bowls found at Praeneste and Kourion. The beast has no fear of humans or respect for their notions of propriety. It represents the worst human qualities housed in a powerful, monstrous form that somehow strongly resembles us.

The Greco-Roman idea of the monkey as an exotic but evil omen also persists in modern times. W.W. Jacobs' short story, *The Monkey's Paw*, first published in 1902, still has a nightmarish quality that reminds us of our limitations. When Sergeant-Major Morris visits the Whites, he tells them the story of the dried monkey's paw he brought back from India. The talisman was said to have been cursed by a fakir with the power to grant three wishes to three separate people with the hope of demonstrating the dangers of trying to alter fate. Morris is the second person to have wished on the paw and claims that he obtained it only after the first man used his third wish to die. Morris throws the object on the fire, but old Mr. White cannot resist the temptation to rescue the paw and wish for a pile of cash. Herbert, Mr. White's son, sits watching the fire after everyone else retires and sees a sinister simian face in the flames. The next day, he dies in a horrible accident at work. His father receives the money he wished for in the form of worker's compensation. Mrs. White uses the second wish to revive her son, but her husband, having learned from his mistake, uses the third wish to return his mutilated son to the grave before the walking corpse can enter their home. The simian face that foretells Herbert's death would be nothing new to the Romans, as evidenced by Suetonius' report of Nero's dream, shortly before his downfall, that his favourite horse had developed the hindquarters of a monkey (*Nero* 46.1).

Recent films, such as *Outbreak*, have played upon more modern fears of deadly diseases carried by monkeys. This 1995 film implausibly portrays a South American capuchin monkey as the carrier of an African virus, similar to Ebola. Another film, *Monkeys Shines* (1988), plays on fears associated with domestic monkeys. This time, a capuchin monkey trained to help a paralyzed man with daily chores, has secretly been programmed to carry out the man's dreams of revenge. The poor monkey goes on a killing spree to avenge its owner, and eventually becomes a threat to him.[34] Although monkeys and apes are capable of violent acts, these portrayals are absurdly negative and affect the modern perception of primates as strongly as Greco-Roman fables affected ancient perceptions.

One must recognize, however, that this study of monkeys is written during a period of cultural change and scientific discovery. The old Greco-Roman speciesist view of human superiority has been severely shaken by the realization that humans and chimpanzees are over 99% genetically similar. There is, in fact, a greater genetic similarity between humans and chimpanzees than there is between chimpanzees and gorillas or orang-utans. So great is the similarity that blood transfusions are possible between chimpanzees, bonobos and humans (Jahme 2000: 9).The use of any animal in medical experiments, particularly our closest living relative, is constantly criticized by animal rights groups. Women now dominate the field of primatology, and new insights into primate behaviour are gaining acceptance. Televised interactions between real apes and humans have led to greater public sympathy for the animals. How many people do not recognize Jane Goodall and the Gombe chimps or Koko, the gorilla who communicates through American sign language, and her keeper, Penny Patterson? Koko's maternal instincts and fondness of kittens have endeared her to humans around the globe. A darker side to primates, though, emerged in 2002 when Goodall's Frodo seized and killed a human child. A debate arose as to whether or not Frodo should be considered a murderer and put to death (BBC 2 2004). Clearly, the line between human and chimpanzee has become thin and blurred. Can Frodo be condemned as a murderer when we have killed hundreds of his species in

[33] It is worth noting that the 2005 remake of King Kong portrays the giant gorilla in a far more sympathetic manner.

[34] In reality, an organization called Helping Hands does train capuchin monkeys to care for quadriplegics. The main criticism against the organization comes from animal rights groups who claim that the animals are unfairly exploited. The monkeys are taken from their mothers in infancy to be raised by human families. After a few years, they are taken from the adopted family and trained to perform simple tasks for their new quadriplegic owners, including combing hair and turning lights on and off. From my few encounters with Helping Hands volunteers, I have learned that the families occasionally become so attached to their capuchin that they are not willing to return the animal to the organization. The change in ownership must also be burden to the monkey.

laboratories, as we test the effects of radiation and seek a cure for AIDS?

Our first-hand observation of primates through the medium of television has done wonders to increase our knowledge of them, but we, as a society, are torn by ideas that have roots in antiquity. We are left with a number of difficult questions: Are we truly superior to apes or are we just another species, perhaps even a third species of chimpanzee? Are we justified in the use of primates in medical experiments or are we killing intelligent, sensitive beings, our own cousins? These questions may take years to resolve, and time is quickly running out for non-human primates as their habitats are destroyed and thousands are killed every year in labs in the US and Europe. I suggest part of the solution resides with the monkeys of the ancient Mediterranean. They hold the keys to understanding some of our oldest prejudices and tell us something about being human in a western society, where Greco-Roman traditions still resonate. Can we not examine these old traditions in the light our newfound awareness to resolve the current monkey puzzle?

Bibliography

ADRADOS, F.R. 1979-1987. *Historia de la fábula greco-latina.* Vol. 1, *Introducción y de los origines a la edad hellenística.* Vol. 2, *la fábula en epoca imperial romana y medieval.* Vol. 3, *Inventario y documentatión de la fábula greco-romana.* Madrid: Editorial de la Universidad Complutense.

ANDERSON, J. 1902. *The Zoology of Egypt: Mammalia.* London: Hugh Reed Ltd.

ANDREU, G. *et al.* 1997. *L'Egypte Ancienne au Louvre.* Paris: Hachette.

ANDREWS, C. 1994. *Amulets of Ancient Egypt.* London: British Museum Press.

ANNAN, D. 1975. *Ape: Monster of the Movies.* Great Britain: Lorrimer Publishing.

ARIS, P.E. nd. *A History of 1000 Years of Greek Vase Painting.* New York: Harry Abrams.

ARMOUR, R.A. 1986. *Gods and Myths of Ancient Egypt.* Cairo: American University in Cairo Press.

ARNST *et al.* 1991. *Ägyptische Museum.* Berlin: Verlag Philipp von Zabern.

ARUZ, J. 1995. Syrian Seals and the Evidence for Cultural Interaction Between the Levant and Crete. In: Matz, F, Ed. *Corpus der Minoischen und Mykenischen Siegel. Beheift 5: Sceaux Minoens et Mycéniens.* Berlin: Mann Verlag, 1-21.

ARUZ, J. 1999. The Oriental Impact on the Forms of Early Aegean Seals. In: Betancourt, P. *et al.*, Eds. *Meletemata: Studies in Aegean Archaeology Presented to Malcolm H. Wiener as He Enters his 65th Year.* Aegeum 20, Liège: Université de Liège, 7-14.

BARGUET, P. 1986. *Textes des Sarcophages Egyptiens du Moyen Empire.* Paris: Les Editions du CERF.

BBC 2. 2004. *The Demonic Ape.* 8 January 2004.

BELER, A.G. de. 2001. *Tutankhamun.* Paris: Molière.

BÉNÉDITE, G. 1911. Scribe et Babouin. *Monuments et Mémoires* 19: 5-42.

BLEEKER, C. 1973. *Hathor and Thoth.* Leiden: Brill.

BIRTACHA, K. and M. ZACHARIOUDAKIS. 2000. Stereotypes in Theran Wall Paintings. In: Sherratt, S., Ed. *The Wall Paintings of Thera: Proceedings from the First International Symposium*, Athens: Thera Foundation, 159-171.

BLACKMAN, A.M. 1915. *The Rock Tombs of Meir,* Vol. 2. London: Egypt Exploration Fund.

BOAS, G., Trans. 1993 [1950]. *The Hieroglyphics of Horapollo.* Bollingen Series 3, Princeton: Princeton UP.

BONACELLI, B.1932. La Scimmia in Etruria. *Studi Etruschi* 6: 341-382.

BOURRIAU, J. 1988. *Pharaohs and Mortals: Egyptian Art in the Middle Kingdom.* Cambridge: Cambridge UP.

BOUZEK, J. 1974. *Graeco-Macedonian Bronzes: Analysis and Chronology.* Prague: Karlova UP.

BOYLAN, P.1922. *Thoth: the Hermes of Egypt.* Oxford: Oxford UP.

BRIER, B. 1994. *Egyptian Mummies: Unraveling the Secrets of Ancient Art.* New York: Quill.

BRIJDER, H.A.G. 1988. Apish Performances in the 6th Cent. BC. In: Christiansen, J. and T. Melander, Eds. *Proceedings of the 3rd Symposium on Ancient Greek and Related Pottery.* Copenhagen: National Museet, 62-70.

BRUNNER-TRAUT, E. 1979. *Egyptian Artists' Sketches: Figured Ostraka from the Gayer-Anderson Collection in the Fitzwilliam Museum, Cambridge.* Istanbul: Nederlands Historisch-Archaeologisch Institut te Instanbul.

BUCHANAN, B. 1966. *Catalogue of Ancient Near Eastern Seals in the Ashmolean Museum,* Vol. 1, *Cylinder Seals.* Oxford: Clarendon Press.

BULTÉ, J. 1991. *Talismans Égyptiens d'Heureuse Maternité.* Paris: CNRS.

CLERMONT-GANNEAU, Ch. 1880. *L'imagerie phénicienne et la mythologie iconique chez les*

Grecs. 1ère partie: la coupe phénicienne de Palestrina. Paris: E. Laroux.

CLINE, E. 1991. Monkey Business in the Bronze Age Aegean. *BSA* 86: 29-42.

CLINE, E. 1994. *Sailing the Wine-Dark Sea: International trade and the Late Bronze Age Aegean.* Oxford: BAR 591.

CNN.com. Sept. 11, 2000. There's No Business (like monkey business). 26 Nov. 2002. http://www.cnn.com/2000/ASIANOW/southest/09/18/monkeybusiness.ap/

COHEN, M.E. 1976. The "Monkey-Letter": A Different Perspective. *Orientalia* 45: 270-74.

COLLON, D. 1982. *The Alalakh Cylinder Seals: A New Catalogue of the Actual Seals Excavated by Sir Leonard Woolley at Tell Atchana and From the Neighbouring Sites on the Syrian-Turkish Border.* Oxford: BAR 132.

COLLON, D. 1995. *Ancient Near Eastern Art.* Berkeley: University of California Press.

CURTIS, L.P. 1971. *Apes and Angels: The Irishmen in Victorian Culture.* New York: Smithsonian Institution.

DASEN, V. 1993. *Dwarves in Ancient Egypt and Greece.* Oxford: Oxford UP.

DAVIS, E. 1986. Youth and Age in the Theran Frescoes. *AJA* 90: 399-406.

DAVIS, T.M. 1908. *The Tombs of Siphtah: The Monkey Tomb and the Gold Tomb.* London: Archibald Constable & Co.

DEMAKOPOULOU, K. 1996. *The Aidonia Treasure: Seals and Jewellery of the Aegean Late Bronze Age.* Athens: Ministry of the Culture of Greece.

DERCHAIN, P. 1952. Bébon, le dieu et les myths. *Révue d'Egyptologie* 9: 23-47.

DOUMAS, Ch. 1985. Conventions Artistiques à Théra et dans la Méditerrannée Orientale à l'Epoque Préhistorique. In: Darcque, P. and J-Cl. Poursat, Eds. *Bulletin de Correspondance Hellénique, Suppl. XI: l'Icononographie Minoenne*, Athens: Ecole française d'Athènes, 29-40.

DOUMAS, Ch. 1992. *The Wall Paintings of Thera*, Athens: Thera Foundation.

DROOP, J.P. 1929. The Laconian Pottery. In: Dawkins, R.M., Ed. *The Sanctuary of Artemis Orthia at Sparta.* London: Macmillan, 52-116.

DUCKWORTH, W.L.H., Trans. 1962. *Galen: On Anatomical Procedures.* Cambridge: Cambridge UP.

DUNBAR, R.I.M. 1977. The Gelada Baboon: Status and Conservation. In: Prince Rainier III and G.H. Bourne, eds. *Primate Conservation.* London: Academic Press, 363-383.

DUNHAM, S. 1985. The Monkey in the Middle. *Zeitschrift für Assyriologie*, Bd.75, II: 234-64.

EIMERL, S. and I. DeVORE. 1965. *The Primates.* New York: Time Inc.

EMERY, W.B. 1965. Preliminary Report on the Excavations at North Saqqara 1964-5. *JEA* 51: 3-8.

EVANS, A. 1921. *The Palace of Minos at Knossos*, Vol. 1, London: Macmillan.

EVANS, A. 1928. *The Palace of Minos at Knossos*, Vol. 2, London: Macmillan.

EVELY, D., Ed. 1999. *Fresco: A Passport into the Past, Minoan Crete Through the Eyes of Mark Cameron.* Athens: British School at Athens.

FA, J.E. and R. LIND. 1996. Population management and viability of the Gibraltar Barbary Macaques. In: Fa, J.E., Ed. *Evolution and Ecology of the Macaque Societies.* Cambridge: Cambridge University Press.

FAIRBANKS, A. 1928. *Museum of Fine Arts, Boston: Catalogue of the Greek and Etruscan Vases*, Vol. 1. Cambridge, MA: Harvard UP.

FAULKNER, R.O., Trans. 1973. *The Ancient Egyptian Coffin Texts*, Vol. 1. Warminster: Aris and Philips Ltd.

FAULKNER, R.O., Trans. 1985 [1972]. *The Ancient Egyptian Book of the Dead.* Andrews, C., Ed. London: British Museum Press.

FAULKNER, R.O., Trans. 1998 [1969] *The Ancient Egyptian Pyramid Texts.* Oxford: Oxford UP.

FISCHER, H.G. 1993. Another Pithemorphic Vessel of the Sixth Dynasty. *JARCE* 30: 1-9.

FJELDHAGEN, M. 1995. *Greco-Roman Terracottas from Egypt.* New York: Carlsberg Glyptotek.

FORBES, P. *et al.*, Eds. 1984. *Encyclopedia of Mammals.* London: George Allen and Unwin.

FOWDEN, G. 1993. *The Egyptian Hermes: A Historical Approach to the Late Pagan Mind.* Princeton: Princeton UP.

FRANKFORT, H. and J.D.S. PENDLEBURY. 1933. *The City of Akhenaten,* Part 2. Oxford: Oxford UP.

FRIEDMAN, F.D. 1995. The Underground Relief Panels of King Djoser. *JARCE* 32: 1-42.

FRIEDMAN, F.D. 1998. *Gifts of the Nile: Ancient Egyptian Faience.* London: Thames and Hudson.

GERMOND, P. 2001. *An Egyptian Bestiary: Animals in Life and Religion in the Land of the Pharaohs*, London: Thames and Hudson.

GILL, M.A.V. 1964. The Minoan Genius. *AM* 79: 430-467.

GOUDSMIT, J and D. BRANDON-JONES. 1999. Mummies of Olive Baboons and Barbary Macaques in the Baboon Catacomb of the Sacred Animal Necropolis at North Saqqara. *JEA* 85: 45-53.

GOUDSMIT, J and D. BRANDON-JONES. 2000. Evidence from the Baboon Catacomb in North Saqqara for a West Mediterranean Monkey Trade Route to Ptolemaic Alexandria. *JEA* 86: 111-119.

GOUDSMIT, J. and R.L. SCHEURLEER. 2000. Monkeys represented in ancient Egypt: image and reality. In: Goudsmit, J., Ed. *Apen in het oude Egypte:*

Archeologie en DNA-onderzoek, Amsterdam: Allard Pierson Museum, 24.

GOPHNA, R. 1993. A Faience Statuette from 'En Besor. *Tel Aviv* 20: 29-32.

GRAFTON, A. 1993. Foreward. In: Boas, G., Trans. *The Hieroglyphics of Horapollo*. Bollingen series 23, Princeton: Princeton UP.

GREEN, L. 2001. Colour Transformations of Ancient Egyptian Pigments. In: Davies, W.V., Ed. *Colour and Painting in Ancient Egypt*. London: British Museum Press, 43-47.

GRIFFITH, F.L. and P.E. NEWBERRY. 1937. *El Bersheh*, Part 2, London: Egypt Exploration Fund.

HAMOTO, A. 1995. *Der Affe in der altorientalischen Kunst.* Münster: Ugarit-Verlag.

HASTINGS, E.A. 1997. *The Sculpture from the Sacred Animal Necropolis at North Saqqara: 1964-76.* London: Egypt Exploration Society.

HERMARY, A.1992. Quelques remarques sur les origins proche-orientales de l'iconographie d'Héraclès. In: Bonnet, C. and C. Jourdain-Annequin, Eds. *Héraclès*. Brussels: Academia Belgica-Ecole française de Rome 129-40.

HIGGINS, R. 1959. *Catalogue of Terracttas in the British Museum,* Vol. 2. London: British Museum Press.

HIGGINS, R. 1967. *Greek Terracottas*. London: Methuen & Co. Ltd.

HIGGINS, R. 1969. *Greek Terracotta Figures*. London: British Museum Press.

HIGGINS, R. 1979. *The Aegina Treasure: An Archaeological History.* London: British Museum Publications.

HIGGINS, R. 1986. *Tanagra and the Figurines*. London: Trefoil Books.

HODEL-HOENES, Sigrid. 2000. *Life and Death in Ancient Egypt: Scenes from the Private Tombs in New Kingdom Thebes.* Warburton, D. Trans. London: Cornell UP.

HOFFMAN, H. 1964. *Some Unpublished Boeotian Satyr Terracottas*. Antike Kunst 7: 67-71.

HOLZBERG, N. 2002. *The Ancient Fable: An Introduction.* Jackson-Holzberg, C., Trans. Bloomington: Indiana UP.

HOOD, S. 1971. *The Minoans: Crete in the Bronze Age*, London: Thames and Hudson.

HOOD, S. 1978. *The Arts in Prehistoric Greece*, New Haven: Yale UP.

HOPKINS, C. 1962. The Aegina Treasure. *AJA* 66: 182-84.

HOULIHAN, Patrick. 1996. *The Animal World of the Pharaohs.* Cairo: American University in Cairo Press.

HOULIHAN, Patrick. 1997. Harvesters or Monkey Business? *GM* 157: 31-47.

HOULIHAN, Patrick. 2001. *Wit and Humour in Ancient Egypt*. London: Rubicon Press.

IKRAM, S. and A. DODSON. 1998. *The Mummy in Ancient Egypt: Equipping the Dead for Eternity.* London: Thames and Hudson.

IKRAM, S. and N. ISKANDER. 2002. *Catalogue Général of Egyptian Antiquities in the Cairo Museum*. Cairo: The Supreme Council of Antiquities Press.

JAHME, C. 2000. *Beauty and the Beasts*. London: Virago.

JANNSEN, R and J. JANNSEN. 1989. *Egyptian Household Animals.* Aylesbury, UK: Shire.

JANSON, H.W. 1952. *Apes and Ape Lore in the Middle Ages and the Renaissance*. London: Warburg Institute.

JENNISON, G. 1937. *Animals for Show and Pleasure in Ancient Rome*. Manchester: Manchester UP.

JOHANSEN, K.F. 1923. *Les vases sicyoniens: Etudes Archéologiques*. Paris: Edward Champion.

KARAGEORGHIS, V. 1994. Monkeys and Bears in Cypriot Art. *Opuscula Atheniensia* XX:6: 63-73.

KARETSOU, Alexandra *et al.* 2000. *Kreti-Aigyptos: Politismoi Desmoi Trion Khilietion.* Heraklion: Heraklion Museum.

KAVANAGH, M. 1983. *A Complete Guide to Monkeys, Apes and Other Primates*. London: Jonathon Cape.

KESSLER, D. and A. el-H. Nur el-DIN. 2002. Inside the Ibis Galleries of Tuna el-Gebel. *EA* 20: 36.

KING, J.E. 1927. *Cicero: Tusculan Disputations*. Loeb Classical Library. London: William Heineman.

KOZLOFF, A. ed. 1981 *Animals in Ancient Art from the Leo Mildenberg Collection.* Cleveland: Cleveland Museum of Art and Indiana UP.

KUMMER, H. 1995. *In Quest of the Sacred Baboon: A Scientist's Journey*. Biederman-Thorson, M.A., Trans. Princeton: Princeton UP.

LACHAUD, R. 1995. *L'Invisible Présence: Les dieux de l'Egypte pharaonique*. Paris: Champollian.

LANGDON, S. 1990. From Monkey to Man: The Evolution of a Geometric Sculptural Type. *AJA* 94: 407-424.

LASSERRE, F. 1959. Bonnard, A. Trans. *Archiloque: Fragments*. Paris: Société d'édition.

LICHTHEIM, M. 1975. *Ancient Egyptian Literature: The Old and Middle Kingdoms*. Berkeley: California UP.

LINDBURG, D., ed. 1980. *The Macaques*. New York: Van Nostrand.

LINSEELE, V. Feb. 2003. Weird Animals From the Elite Cemetery. *Archaeology's Interactive Dig.* 10 June, 2003.http://www.archaeology.org/interactive/hierakonpolis/animals.html.

LISSARRAGUE, F. 1990. The Sexual Life of Satyrs. In: Halper, D.M. *et al.*, Eds. *Before Sexuality: The Construction of Erotic Experience in the Ancient World.* Princeton: Princeton UP, 52-81.

LISSARRAGUE, F. 1993. On the Wildness of Satyrs. In: Carpenter, T.H. and C.A. Faraone, Eds. *Masks of Dionysus*. New York: Cornell UP, 207-20.

LISSARRAGUE, F. 1997. L'homme, le singe et le satyre. In: Cassin, B. and J.-L. Labarrière, eds. *L'Animal dans l'antiquité*. Paris: Centre National de la Recherche Scientifique: 455-472.

LISSARRAGUE, F. 2000. Aesop, Between Man and Beast: Ancient Portraits and Illustrations. In: Cohen, B. ed., *Not the Classical Ideal: Athens and the Construction of the Other in Greek Art.* Leiden: Brill: 132-149.

LLOYD-JONES, H., Trans.1974. *Simonides: Females of the Species: Simonides on Women.* Park Ridge, NJ: Noyes Press.

LORTET, L. and M.C. GAILLARD. 1907. *La Faune Momifiée de l'Ancienne Egypte et Recherches Anthropologiques*, Vol. 3. Lyon: Librairie Henri George.

McDERMOTT, W.C. 1935. The Ape in Greek Literature. *Transactions of the American Philological Association* 66: 165-76.

McDERMOTT, W.C. 1936. The Ape in Roman Literature. *Transactions of the American Philological Association* 67: 148-67.

McDERMOTT, W.C. 1938. *The Ape in Antiquity.* Baltimore: Johns Hopkins Press.

MALEK, J. 1997. *The Cat in Ancient Egypt*. Philadelphia: University of Pennsylvania Press.

MALLOWAN, M.E.L. 1947. Excavations at Brak and Chagar Bazar. *Iraq* 9: 1-87.

MANNICHE, L. 1997. *Sexual Life of the Ancient Egyptians*. London: Kegan Paul Intl.

MANNICHE, L. 1999. *Sacred Luxuries: Fragrance, Aromatherapy and Cosmetics in Ancient Egypt.* London: Opus Publishing.

MARINATOS, N. 1984a. *Art and Religion in Thera*, Athens: Mathioulakis.

MARINATOS, N. 1984b. The Date Palm in Minoan Iconography. *Opuscula Atheniensia* 15: 115-22.

MARINATOS, N. 1987a. An Offering of Saffron to the Minoan Goddess of Nature. *Boreas* 15: 123-32.

MARINATOS, N. 1987b. The Monkey in the Shrine: A Fresco Fragment from Thera. *Eilapine*: 417-19.

MARKOE, G. 1985. *Phoenician Bronze and Silver Bowls from Cyprus and the Mediterranean.* Berkeley: University of California Press.

MARQUAND, A. 1887. A Silver Patera from Kourion. *AJA* 3: 322-37.

MELLINK, M.J. 1987. Anatolian Libation Pourers and the Minoan Genius. In: Farkas, A.E. *et al.*, Eds. *Monsters and Demons in the Ancient and Medieval Worlds.* Mainz: Verlag Philipp von Zabern: 65-72.

MENDLESON, C. 1983. More Monkey Business. *Anatolian Studies* 33: 81-3.

MERCER, S.A.B., Trans. 1952. *The Pyramid Texts.* New York: Green and Co.

MESKELL, L. and R.A. JOYCE. 2003. *Embodied Lives: Figuring Ancient Maya and Egyptian Experience.* London: Routledge.

MILLER, F.J., Trans. 1964. *Ovid's Metamorphoses*. Vol. 2, Loeb Classical Library, Cambridge, MA: Harvard UP.

MOLLARD-BESQUES, S. 1954. *Musée du Louvre: Catalogue raisonné des figurines et reliefs en terre-cuite grecs, étrusques, et romains.* I. *Epoques prehellénique, géométrique, archaïque, et classique.* Paris: Editions des Musées Nationaux.

MOLLARD-BESQUES, S. 1986. *Musée du Louvre: Catalogue raisonné des figurines et reliefs en terre-cuite grecs, étrusques, et romains.* IV.2 *Epoques hellénistique et romaine, cyrénaique, Egypte ptolémaique et romaine, Afrique du Nord et Proche-Orient.* Paris: Editions des Musées Nationaux.

MORGAN, L. 1988. *The Miniature Wall Paintings of Thera.* Cambridge UP.

MORGAN, L. 1995. Frontal Face and the Symbolism of Death in Aegean Glyptic. *CMS Beiheft* 5: 135-150.

Museum of Fine Arts, Boston. 1982. *Egypt's Golden Age: The Art of Living in the New Kingdom 1558-1085 BC.* Catalogue of the Exhibition, Boston: Museum of Fine Arts.

NOWAK, R.M. 1999. *Walker's Primates of the World.* Baltimore: Johns Hopkins University Press.

OIKONOMIDES, Al.-N., Trans. 1977. *Hanno the Carthaginian: Periplus or Circumnavigation [of Africa].* Chicago: Ares.

OSBORN, D.J. and J. OSBORNOVA. 1998. *The Mammals of Ancient Egypt*. Warminster: Aris and Phillips, Ltd.

PALLOTINO, M., Ed. 1980. *Il Museo Nazionale Etrusco di Villa Giulia.* Rome: Quasar.

PERRY, B.E., Trans. 1965. *Babrius and Phaedrus*. Loeb Classical Library. Cambridge and London: Harvard University Press.

PERRY, B.E., Trans. Ed. 1952. *Aesopica: A Series of Texts Relating to Aesop or Ascribed to Him or Closely Connected with the Literary Tradition That Bears His Name. Collected and Critically Edited, in Part Translated from Oriental Languages, with a Commentary and Historical Essay*. Vol. 1, *Greek and Latin Texts*. Urbana: University of Illinois Press.

PINCH, G. 1994. *Magic in Ancient Egypt*. London: British Museum Press.

POULIANOS, A. 1972. The Discovery of the First Known Victim of Thera's Bronze Age Eruption. *Archaeology* 25: 229-230.

POWELL, M.A. 1978. Ukubi to Mother … The Situation is Desperate. *Zeitschrift für Assyriologie* 68: 169-95.

PRESTON, L.E. 1975. Four Boeotian Ape Figurines from the J. Paul Getty Museum. *J. Paul Getty Museum Journal* 2: 121-6.

PRESTON-MAFHAM, R. and K. 1999. *Primates of the World.* London: Blandford.

RACKHAM, H., Trans. 1940. *Pliny: Natural History*. Loeb Classical Library. Cambridge, MA: Harvard UP.

RATNAGAR, S. 1981. *Encounters: The Westerly Trade of the Harappa Civilization*. Delhi: Oxford University Press.

REBUFFET-EMMANUEL, D. 1967. Singes de Mauritanie Tingitane et d'Italie: Reflexions sur une analogie iconographique. *Studi Etruschi* 35: 633-644.

RICHTER, G.M.A. 1930. *Animals in Greek Sculpture: A Survey*. Metropolitan Museum of Art. London: Oxford University Press.

ROBERTS, A. 2000. *My Heart My Mother: Death and Rebirth in Ancient Egypt*. Rottingdean, UK: Northgate.

ROBINSON, D.M. 1934. *CVA* Baltimore 1 (USA 4). Cambridge, MA: Harvard UP.

ROMANO, James. 1989. *The Bes Image in Pharaonic Egypt*, Vol. I. Ann Arbor: UMI Dissertation Services.

RUTTEN, M. 1938. Les animaux à attitude humaine dans l'art de l'Ancienne Mésopotamie. *Revue des Etudes Sémitiques*: 97-119.

SALAMAN, C. *et al*, Trans. 1999. *The Way of Hermes: The Corpus Hermeticum*. London: Duckworth.

SAMSON, J. 1978. *Amarna: City of Akhenaten and Nefertiti*. Warminster: Aris and Phillips.

SANTOLINI, G. 1984. Thot, le babouin et le palmier doum. *Mélanges Adolphe Gutbub*. Montpellier: Institut d'Egyptologie-Université Paul Valéry, 211-18.

SCHOFIELD, A.F., Trans. 1959. *Aelian: On Animals*, Vol. 3. Loeb Classical Library. Cambridge, MA: Harvard UP.

SCHWEITZER, B.1971. *Greek Geometric Art*. London: Phaidon.

SHAW, I. and P. NICHOLSON. 2002. *The British Museum Dictionary of Ancient Egypt*. London: British Museum Press.

SHERKOVA, T. 2000. Seven Baboons in One Boat: The Interpretation of Iconography in the Context of the Cult Belonging to the Temple at TellIbrahim Awad. Hawass, Z. Ed. *Egyptology at the Dawn of the Twenty-first Century: Proceedings of the Eighth International Congress of Egyptologists*. Cairo: The American University in Cairo Press: 504-08.

SHERRATT, S, Ed. 2000. *The Wall Paintings of Thera: Proceedings from the First International Symposium*, Athens: Thera Foundation.

SIMPSON, W.K. 1973. *The Literature of Ancient Egypt*. New Haven: Yale University Press.

SMITH, H.S. 1974. *A Visit to Ancient Egypt: Life at Memphis and Saqqara (c.500-30 BC)*. Warminster: Aris and Phillips.

SOUTHWICK, C.H. and M.F. SIDDIQI. 1977. Population Dynamics of Rhesus Monkeys in Northern India. In: Prince Rainier III and G.H. Bourne, Eds. *Primate Conservation*. New York: Academic Press, 339-62.

SPENCER, A.J. 1993. *The Rise of Civilisation in the Nile Valley*. London: British Museum Press.

SPERLING, S. 1991. Baboons with Briefcases vs. Langurs in Lipstick: Feminism and Functionalism in Primate Studies. In: di Leonardo, M., Ed. *Gender at the Crossroads of Knowledge: Feminist Anthropology in the Postmodern Era*. Berkeley: University of California Press, 204-34.

SWINDLER, M.H. 1929. *Ancient Painting*. New Haven: Yale UP.

TAMBIAH. J. 1985. Animals Are Good to Think and Good to Prohibit. In: Tambiah, S.J. ed., *Culture, Thought and Social Action: An Anthropological Perspective*. Cambridge, MA: Harvard UP, 169-211.

TAMVAKI, A. 1973. Some Unusual Mycenaean Terracottas. *BSA* 68: 207-65.

TERRACE, E. 1966. Blue Marble Plastic Vessels. *JARCE* 5: 57-63.

Te VELDE, G. 1988. Some Remarks on the Mysterious Language of Baboons. *Funerary Symbols and Religion*. Kamstra, J.H. *et al*., Eds. Kampen: J.H. Kok, 129-136.

THITE, G.U. 1983. Indian Fable. In: Adrados, F.R. and O. Reverdin, Eds., *La Fable: Huit exposés suivis de discussions*. Entretiens sur l'antiquité classique, 30. Geneva: Reverdin and Grange, 33-60.

TOYNBEE, J.M.C. 1996 [1973]. *Animals in Roman Life and Art*. Baltimore: Johns Hopkins UP.

URE, P.N. 1934. *Aryballoi and Figurines in from Rhitsona in Boeotia*. London: Cambridge University Press.

VALLOGGIA, Michel. 1993. Une Coupe à décor thériomorphe provenant de Balat. *BIFAO* 93: 391-402.

Van BUREN, E.D. 1930. *Clay Figurines of Babylonia and Assyria*. New Haven: Yale University Press.

Van GENNEP, A. 1960 [1909]. The Rites of Passage. Vizedom, M.B., Trans. London: Routledge.

VANDIER d'ABBADIE, J. 1936. *Catalogues des ostraca figures de Deir el Médineh*. Fasc. 1, Cairo: Imprimerie de l'Institut Français d'Archéologie Orientale.

VANDIER d'ABBADIE, J. 1937. *Catalogues des ostraca figures de Deir el Médineh*. Fasc. 2, Cairo: Imprimerie de l'Institut Français d'Archéologie Orientale.

VANDIER d'ABBADIE, J. 1964. Les Singes Familiers I: l'Ancien Empire. *Revue d'Egyptologie* 16: 147-177.

VANDIER d'ABBADIE, J. 1965. Les Singes Familiers II: le Moyen Empire. *Revue d'Egyptologie* 17: 177-88.

VANDIER d'ABBADIE, J. 1966. Les Singes Familiers dans l'Ancienne Egypte III: le Nouvel Empire. *Revue d'Egyptologie* 18:143-201.

VANDIER d'ABBADIE, J. 1972. *Catalogue des objets de toilette égyptiens au Musée du Louvre*. Paris: Musée du Louvre.

VANSCHOONWINKEL, J. 1990. Animal Representations in Theran and Other Aegean Arts. In: Hardy, D.A. and C. Doumas, Eds. *Thera and the Aegean World* III, Vol. 1, London: Thera Foundation, 143-201.

WALKER, A., Ed. 1996. *Animals in Art from the Leo Mildenberg Collection, III*. Mainz: Philipp von Zabern.

WEINGARTEN, J. 1991. *The Transformation of the Egyptian Taweret into the Minoan: A Study in Cultural Transmission in the Middle Bronze Age*. SIMA 87, Partille: Paul Astrom.

WHITEHOUSE, H. 2001. *Ancient Mosaics and Wallpaintings*. London: Harvey Miller.

WILKINSON, T. 1999. *Early Dynastic Egypt*. London: Routledge.

WILKINSON, R. 1999. *Symbol and Magic in Egyptian Art*. London: Thames and Hudson.

WILKINSON, R. 2000. *Reading Egyptian Art: A Hieroglyphic Guide to Ancient Egyptian Painting and Sculpture*. London: Thames and Hudson.

WINTER, F. 1903. *Die Typen der figürlichen Terakotten*, Vol. 1. Berlin: Verlag von W. Spemann.

WINTER, I.J. 1983. The Program of the Throneroom of Assurnasirpal II. In: Harper, P.O. and H. Pittman, Eds. *Essays on Near Eastern Art and Archaeology in Honor of Charles Kyle Wilkinson*. New York: Metropolitan Museum of Art, 15-29.

WOOLLEY, C.L. 1934. *Ur Excavations*, Vol. 2. *The Royal Cemetery: a Report on the Predynastic and Sargonid graves excavated between 1926 and 1931*. London: British Museum Press.

XANTHOUDIDES, S. 1971 [1924]. *The Vaulted Tombs of the Mesara*. Westmead, England: Gregg International Publisher's Ltd.

XÉNAKI-SAKELLARIOU, A. 1958. *Les Cachets Minoens, EtCrét* X. Paris: P. Geuthner.

YOYOTTE, J. 1991. Preface. Bulté, J. Ed. *Talismans Égyptiens d'Heureuse Maternité*. Paris: CNRS.

www.ingramcontent.com/pod-product-compliance
Lightning Source LLC
LaVergne TN
LVHW071220210926
848482LV00015B/859